Published by Aperitifs Publishing Company
Santa Rosa, California
707-523-1611
johncburton@msn.com

ISBN: 978-1-7324530-1-2
Library of Congress number: 2019902092

Copyright: March 2019
Written by:
John C. Burton
John Louder

Printed in the United States of America

All rights reserved. No part of this book may be reproduced or transformed in any form or by any means, electronic or mechanical, including photocopying, recording or by any information storage and/or retrieval system without permission in writing from the author or publisher.

Every attempt has been made to provide accurate information on the following subjects.

johncburton@msn.com
707-523-1611

ACKNOWLEDGEMENTS
FRONT COVER IMAGE

The J. Franetta bottle on the cover is courtesy of Rick Siri collection and the Tamalpais Mineral Water label courtesy of Newall Snyder's collection. Both the Frey & Lenz bottle and A. Timony label are a part of John C. Burton collection.

REAR COVER IMAGE

The Ell-Ell Whiskey and San Rafael Soda Works P. & B. Prop's images are from Jeff Wichmann's American Bottle Auctions catalog of Sacramento. The Eddie Golden paper-maché covered half-pint is from John Louder's collection.

- John Louder who co-authored this endeavor and supplied many bottles from his personal collection and spent many hours at the Sonoma County Library on 3rd Street in Santa Rosa.
- Rick Siri who graciously allowed us to photograph his extremely rare J. Franetta bottle. This bottle is one of the most sought-after Marin bottles.
- Richard Siri for allowing us to photograph his bottles and hosting the Northwestern Bottle Club meetings every month at his facility.
- Dr. Thomas Jacobs who reviewed our outline and made major corrections to the book. Also, for his contributions to the Lang and Frey & Company section and especially the photograph of his Frey & Co. paper label bottle.
- Dan Brown who opened his house to John Louder and me and allowed us to photograph his great collection and his contribution of the Klammer & Malz Petaluma and San Rafael bottle, Mason Sausalito Gingerale crown top sodas, the Senate and J. E. Brady pumpkinseeds to this book.
- Newall Snyder who always shares items from his personal museum. Without his contributions our books would never be complete.
- Chuck Ingraham for his contribution of the Henry C. Gieske Dandy flask.
- Eric McGuire for his photo of Salvadore Grandi's whiskey bottle and information regarding Salvadore.
- Neal Austinson for use of his Edmond Meyer Coca Cola Business card from his collection.
- San Rafael Historical Society for images and most importantly to Carol Acquaviva Librarian /Digital Archivist in the California Room at the Marin County Free Library for her assistance in finding items to research.
- And to the on-line California Digital Newspaper where we found related items in the San Francisco Call, the Marin County Journal and the Marin County Tocsin. In addition, we thank Sacramento Daily Union for their reference as well. If you do research, we highly recommend the California Digital Newspaper as your best source.
- Sarah Everson who has assisted me on our publications;
    BOTTLES, TOKENS &HISTORY OF SONOMA COUNTY
    GRACE BROS. BREWERIES, HISTORY & MEMORBILIA
    LAKE, NAPA, SONOMA, MENDOCINO, SOLANO, MARIN and HUMBOLDT MINERAL & HOT SPRINGS

John C. Burton
John Louder

TABLE OF CONTENTS

List of Chronological order of owners and San Rafael Brewery site image	Page 1
Images of Brewery and Stables	Page 2
Austin J. McLellan & Rufus A. Roscoe founders & developers of San Rafael Brewery	Page 3
Gustave Lambert short term owner	Page 4
Henry Boyen & Fritz Goerl partners	Page 5
San Rafael Brewery & Gardens advertisement and Brewery letterhead	Page 6
Henry Boyen & Fritz Goerl photo with staff	Page 7
Brewery advertisement & billhead	Page 8
Photo of W. O. Smith maintenance man repairing wagon	Page 9
Jacob Blum driving delivery wagon	Page 10
Advertisement for Henry Boyen's Pioneer Cash Grocery store	Page 11
Chronological order of owners and rendering of San Rafael Bottling plant	Page 12
Notice of co-partnership; Frey & Lenz	Page 13
Frey & Lenz advertisement	Page 14
Additional Frey & Lenz advertisements	Page 15
Frey & Lenz beer bottle	Page 16
Frey & Lenz paper label cabernet wine label	Page 17
Dissolution of Frey & Lenz partnership	Page 18
Declaration by Frey & Company	Page 19
Declaration of property by Frey & Company	Page 20
Frey & Company advertisements Agents for Wieland's Lager	Page 21
Images of Frey & Co. applied top bottles	Page 22
Image of Frey & Co. crown top bottle	Page 23
1910 Paper label Fredericksburg beer bottle Frey & Co. Sole Bottlers	Page 24
Listing of products sold by Frey & Co.	Page 25
I.W. Harper whiskey sold by Frey & Co.	Page 26
Wines sold by Frey & Co.	Page 27
Wine advertisement wholesale and retail	Page 28
Old Kentucky Rye whiskey	Page 31
Fredericksburg advertisement and bottle	Page 32
Images of soda and mineral water bottles sold by Frey & Co.	Page 33
Photos of Franetta wine & liquor store	Page 34
Franetta advertisements	Page 35
Fredericksburg advertisements and bottle images	Page 36
Advertisement regarding wholesale wine & liquor	Page 37
John Franetta bottle in Rick Siri collection & advertisement from Jacobs collection	Page 38
George Nowell & George Franetta new partners	Page 39
Dissolution of partnership of Nowell & Franetta	Page 40
Notice of co-partnership Otto, Leonard & August Lang with Cohoon & Nowell	Page 41
Continuation of partnership article	Page 42
Dissolution of partnership removing George Nowell	Page 43
Dissolution of partnership removing W. O. Cohoon	Page 44
Lang Bros. billing statement and Philadelphia & Lang S. F. Bottle	Page 45
Outline of Lang & Company	Page 46
Marin County Club Colonel Stubbs	Page 47

S. A. Pacheco Marin County Club whiskey distributor	Page 48
Three Marin County Club whiskey bottle images	Page 49
Salvatore Grandi of Point Reyes Station	Page 50
Grandi's Special Whiskey bottle image	Page 51
Notice of Jack Brady opening a saloon in Mill Valley	Page 52
J. E. Brady Sequoia pumpkinseed bottles	Page 53
J. E. Brady Sequoia Mill Valley Cal. Fifth	Page 54
Henry G. Gieske Family Liquor Store bottle	Page 55
The Senate S. Kerrigan San Rafael bottle	Page 56
A. Timony bourbon label & notice of opening saloon	Page 57
Notice of sale of A. Timony's saloon	Page 58
Eagle Bar paper-mâché ½ pint bottle	Page 59
Golden Eagle Hotel photo	Page 60
Golden Eagle Hotel paper-mâché ½ pint bottle	Page 61
Zopf's Wine gardens trade card	Page 62
Chapter 2 Marin County Soda Works	Page 63
San Rafael Soda Works Joseph Kappenman bottle	Page 64
Kappenman's saloon inside Central Hotel	Page 65
San Rafael Soda Works advertisement	Page 66
San Rafael Soda Works Sylvan Provencal & Alphonse Bresson	Page 67
Marin Soda & Iron Phosphate Works Martin Petersen & chronological list of owners.	Page 68
Marin Directory listing of Martin Petersen's products & Petersen photo	Page 69
M. Petersen Marin Soda Works advertisement in German	Page 70
Martin Petersen hutch bottle	Page 71
Additional images of Martin Petersen bottles	Page 72
Petersen notice of manufacturing	Page 73
Petersen trademarks	Page 74
Marin County Bottling Works	Page 75
Klammer & Malz hutch bottle	Page 76
Klammer & Malz crown top bottles	Page 77
Klammer & Malz seltzer bottle	Page 78
Notice of dissolution of Klammer & Malz	Page 79
Buffalo Bottling Works	Page 80
Mt. Tamalpais Water Analysis	Page 81
Buffalo Soda Works directory back cover	Page 82
Tamalpais Natural Mineral water bottle	Page 83
Borello Advertisements	Page 84
Borello Advertisements	Page 85
Buffalo hutch bottles	Page 86
Borello crown top bottles & Buffalo box	Page 87
Borello seltzer bottle	Page 88
Notice of Borello Brothers purchasing Petersen trademark, bottles & copper kegs....	Page 89
Chronological order of owners of San Anselmo Bottling Company	Page 90
San Anselmo hutch bottles	Page 91
San Anselmo crown top bottles	Page 92
San Anselmo seltzer bottles	Page 93
Meyer's Beverage Company	Page 94

Meyer's paper label sodas	Page 95
Meyer's painted label bottle and crown top opener	Page 96
Meyer's painted label bottle	Page 97
Meyer's embossed crown top bottles	Page 99
Meyer's quart size bottles	Page 100
Meyer's seltzer bottles	Page 101
Meyer's Coca Cola business card & empty bottles advertisement	Page 105
Notice that Meyer's is operating under laws of California using following trademarks Of Marin Soda Works, Marin Bottling Works, E. Malz, Klammer & Malz, & Petersen.	Page 106
Meyer's advertisement regarding Coca Cola & Coca Cola advertisement	Page 108
Mason American Distillery plant	Page 109
Sheriff's Sale against Mason Malt Whiskey Distilling Company	Page 110
Lawsuit against Mason for pirating labels	Page 111
Lawsuit against Mason for foul smells surrounding Pine Station	Page 112
Mason announces that smell will end & also announces closing of distillery	Page 113
Mason Brothers & John Buck plan to reopen plant	Page 114
Federal agents visit Mason Plant	Page 115
Mason plant cleared to reopen	Page 116
Ernest Hopkins tells of making Dry Ice	Page 117
DeWitt & John Mason sell distillery and liquidate all trademarks and equipment	Page 118
Clint Mason sued by Lawyer Louis Pistolesi for legal fees	Page 119
Mason hutch bottle	Page 120
Mason crown top bottles with embossing	Page 121
Mason crown top embossed bottles	Page 122
Mason seltzer bottles	Page 123
San Rafael – San Francisco Fortuna Water bottle	Page 125
Recommended Books	Page 126
Show Chair Persons	Page 128

Tamalpais Mineral Water label Newall Snyder collection

CHAPTER 1
SAN RAFAEL BREWERY

- 1870 - 1871   Austin McLellan & Rufus A. Roscoe
- 1871 – 1872   Gustave Lambert
- 1872 – 1877   Henry Boyen
- 1877 – 1884   Henry Boyen & Fritz Goerl
- 1884 – 1886   Fritz Goerl & Louis Graeber
- 1886 – 1899   Fritz Goerl & Company
- 1899 – 1905   Fritz Goerl & Son George Goerl
- 1905 – 1907   Thomas H. Chapman, Claude H. Huckle & Elbert J. Fryman
- 1907 – 1920   Chapman, Huckle, Fryman, Ernest E. Reed & Percy
- 1920   Prohibition

MAIN HOUSE AT THE BREWERY SITE
San Rafael Historical Society photo

VIEW OF THE BREWERY & STABLES
San Rafael Historical Society photo

Brewery operation on the right. Notice two teams of horse and wagons heading towards the brewery and one team leaving the brewery on the main road.

Newall Snyder Collection

DEVELOPERS OF SAN RAFAEL BREWERY

San Rafael Brewery was built in 1870 by Austin J. McLellan and Rufus A. Roscoe who opened the Brewery May 1871. Located on Greenwood Avenue these two Canadian men operated the brewery until September 1871 when they ran into financial trouble selling to Gustave Lambert.

Lambert died within a few months of purchase and the brewery was purchased by Henry Boyen who operated the brewery until 1877 when he took in Fritz Goerl as a partner.

Boyen had been a brewer in Germany and then at the Albany Brewery in San Francisco for five years. Goerl had been a brewer in Alaska prior to moving to San Rafael. In 1880 Boyen sold his share to Louis Grabber and purchased a saloon in San Rafael passing away in 1890.

The new partnership only lasted until 1884 when Goerl purchased Grabbers share operating as sole proprietor until 1898 when he took in his son George as a partner.

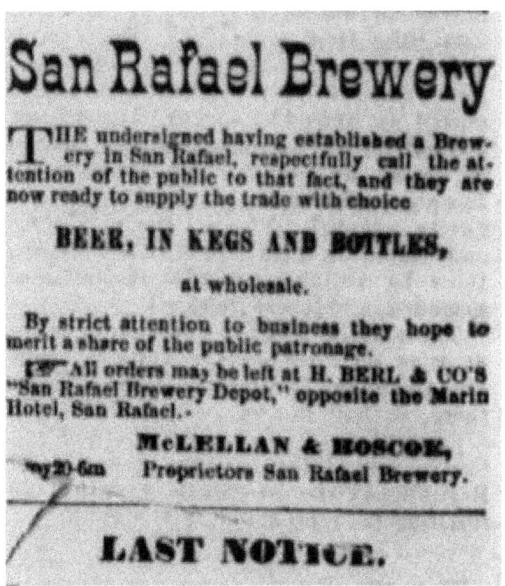

**This is the last brewery ad placed in the Marin Journal by McLellan & Roscoe
Marin Journal May 20, 1871**

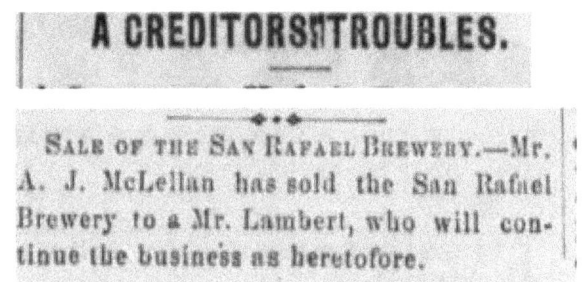

**Notice of San Rafael Brewery sale to Gustave Lambert
Marin Journal September 29, 1871**

SAN RAFAEL BREWRY,

GREENWOOD AVENUE, SHORT'S ADDITION, SAN RAFAEL.

G. Lombert & Co., Proprietors.

HAVE constantly on hand and will deliver to customers a choice article of

BEER and ALE.

The proprietors being practical Brewers, the public may rely upon getting a good article.

m4-tf

Gustave Lambert advertisement.
Notice that they spelled his last name Lombert

On Sunday last Gustave Lombard, proprietor of the San Rafael Brewery, died suddenly at his residence in in this place. On the morning of the day on which he died he arose and was about to proceed to his business as usual, when he complained of a dizziness in his head and a pain in his left side. In a few minutes thereafter he fell to the floor, when a physician was immediately summoned, and in a few minutes after his arrival Lombard breathed his last.

Notice that Gustave Lambert died.
Unfortunately, they listed his last name as Lombard.
Sacramento Daily Union July 31, 1872

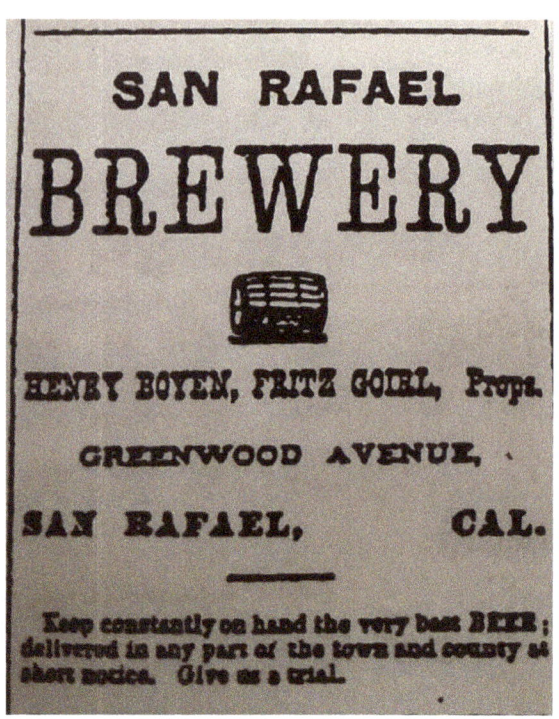

Henry Boyen & Fritz Goerl
Boyen and Goerl remained partners from 1877 until 1884.

Goerl had many troubles with revenue agents regarding omitting tax stamps on his kegs. Without tax stamps the government would not be collecting tax money. At times the issue was blamed on employees who were lax or when driver John Pachner would remove the tax stamps after delivering the beer. The stamps would be used again and again eliminating payment of tax on goods sold. In 1897 the government seized the brewery and fined Goerl $2,501.50.

Apparently Pachner continued to remove tax stamps as in 1899 the brewery was seized a second time by the government for the same offense. Goerl paid the government $1,000 penalty and $1,309 for evading taxes.

In 1897 they were fined $2,501.50 and in 1899 fined $1,000 and $1,309 for evading payment of taxes.

FRITZ GOERL & COMPANY

SAN RAFAEL BREWERY AND GARDENS SUMMER RESORT

SAN RAFAEL BREWERY AND GARDENS SUMMER RESORT.

Beer and Porter, by the Keg or Bottle,
Delivered in any part of San Rafael Township Free of Charge.

FRITZ GOERL, Proprietor.

Greenwood Avenue and Madrone Sts., South end of Town, SAN RAFAEL.

1890 advertisement from rare Marin Directory Dr. Thomas Jacobs collection.

September 1, 1897 statement Newall Snyder collection.

SAN RAFAEL STEAM BREWERY
STAFF PHOTO

Their products were lager, porter and steam beer sold in bottles and kegs.
Henry Boyen on left and Fritz Goerl on far right.
Photo San Rafael Historical Society collection.

SAN RAFAEL BREWERY ADVERTISEMENT AND LETTERHEAD

Marin Journal December 23, 1886

The Goerl's sold the brewery in January 1906 to Elbert J. Fryman, Thomas H. Chapman and Claude H. Huckle. They not only brewed San Rafael beer but became agents for Acme beer.

San Rafael Historical Society collection.
San Rafael brewery also distributed Acme Beer

The three men incorporated and operated the brewery until Prohibition in 1920.

March 16, 1906 Fritz Goerl was riding with his wife Josephine (Josie) in their horse drawn buggy. The team of horses was spooked running away hitting a eucalyptus tree. Fritz fractured his skull and died within hours.

George Goerl left San Rafael and managed the Palace Brewery in Alameda from 1907 until 1910. As with many breweries, San Rafael Brewery closed with Prohibition never to open again.

W. O. SMITH
MAINTENENCE

Photo San Rafael Historical Society collection.
W. O. Smith, carriage maker and general blacksmith posing under the wagon while two co-workers stand by.

JACOB BLUM
DELIVERY MAN

Photo San Rafael Library Collection

Jacob Blum was a driver for the brewery. When he left the San Rafael Brewery, he opened a saloon on the corner of Fourth & Lincoln streets in the late 1880's.

In 1897 he became the Master Brewer for Grace Bros. Santa Rosa. With his passing his son Oscar became the Master Brewer for Grace Bros.

HENRY BOYEN'S NEW VENTURE

After dissolving his partnership with Fritz Goerl; Henry Boyen opened Pioneer Cash Grocery store two blocks from the brewery.

Pioneer CASH GROCERY,
CORNE
FOURTH AND F STREETS,
SAN RAFAEL.
H. BOYEN, PROPRIETOR

CHOICE
STAPLE AND FANCY GROCERIES,
Of All Descriptions. Also
Provisions,
Produce,
Teas Coffees,
Spices, Etc

SOLE AGENT for the brated
GENUINE LAGER BEER
OF THE NATIONAL BREWING CO.,
Which I Can Supply in Quantities to Suit

SMALL PROFITS. STERLING GOOD

☞Please call and get my prices before going elsewhere.

Marin Journal
August 4, 1897

SAN RAFAEL BOTTLING COMPANY
FREY & COMPANY

- 1897 – 1897   Franz Frey & Richard Lenz
- 1897 – 1920   Franz Frey
- 1920 - 1933   Prohibition

Franz Frey immigrated to the United States from Glasgow April 30, 1891 coming to San Rafael and opening a wholesale liquor store at 901-905 Fourth Street. On May 27, 1897 he partnered with Richard Lenz. As with many shop owners of the time Frey's residence was above the business. Lenz had been an employee of the Lang Brothers in San Francisco.

RENDERING OF BOTTLING PLANT
Corner 4th & D Streets

Marin County Tocsin
Post partnership advertisement
July 1, 1899

NOTICE OF CO-PARTNERSHIP
FRANZ A. FREY & RICHARD LENZ

NOTICE OF CO-PARTNERSHIP.

STATE OF CALIFORNIA, } ss.
County of Marin.

WE, THE UNDERSIGNED, DO HEREBY certify that we are co-partners doing business in the City of San Rafael, in the County of Marin, State of California, under the firm name of "FREY & LENZ," and that the names in full of all the members of said partnership are as follows:

Franz A. Frey, residing at the City of San Rafael, County of Marin, State of California.

Richard Lenz, residing at the City of San Rafael, County of Marin, State of California.

In witness whereof we have hereunto set our hands this 11th day of January, 1897,

FRANZ A. FREY,
RICHARD LENZ.

STATE OF CALIFORNIA, } ss.
County of Marin.

On this 11th day of January, in the year A. D. 1897, before me James W. Cochrane, a Notary Public in and for the said County of Marin, personally appeared Franz A. Frey and Richard Lenz, known to me to be the persons whose names are subscribed to the within instrument, and they acknowledged to me that they executed the same.

In witness whereof, I have hereunto set my hand and affixed my official seal the day and year in this certificate first above written.

JAMES W. COCHRANE,
Notary Public,
In and for Marin County, State of California.

Marin County Tocsin
January 11, 1897

FREY & LENZ ADVERTISEMENT

Messrs. Frey & Lenz whose advertisment appears in the columns of this paper have purchased the business of Lang Bros and will continue at the old stand, corner of 4th and D streets. Years of experience in the wholesale and retail liquor business, have given to the firm a prominent standing as purveyors of pure and unadulterated articles on this coast. The trade, families, and the public generally will find in ordering from Frey & Lenz first-class goods at moderate prices.

Marin Journal
April 12. 1897

FREDERICKSBURG BREWERY.

Frey & Lenz   Telephone Main 10.

Wholesale **WINE AND LIQUOR MERCHANTS**

Goods delivered free to any part of San Rafael or Ross Valley.

COR. FOURTH AND D STREETS.

ORDERS BY MAIL OR TELEPHONE Promptly attended to.

Marin Journal
August 12, 1897

FREY & LENZ ADVERTISEMENTS

Marin County Tocsin advertisement
February 6, 1897

Notice the telephone number change from 10 to 35

FREY & LENZ BEER
EXTREMELY RARE FREY & LENZ BOTTLE
(Two known)

PROPERTY OF
FREY & LENZ

(P C G W)
Pacific Coast Glass Works

Beer was purchased in kegs by Frey & Lenz from Fredericksburg brewery in San Jose and Lang Brothers in San Francisco then transferred to embossed bottles with their name and city on the face of the bottles.

A very common practice by bottlers on the west coast; F. O. Brandt of Healdsburg, Hudson & Palmer of Santa Rosa, Zimmerman Brothers of Sebastopol, & Louis Schmidt of Pataluma as examples.

FREY & LENZ
CABERNET WINE BOTTLE

Frey & Lenz paper label Cabernet wine label.
Bottle Newall Snyder collection.

DISSOLUTION OF PARTNERSHIP
November 1, 1897

Dissolution of Copartnership.

THE FIRM HERETOFORE EXISTING UNder the name and style of "Frey & Lenz," doing business in the City of San Rafael, Marin County, State of California, is this day dissolved by mutual consent, Richard Lenz withdrawing from said copartnership.

Said business will be hereafter carried on under the firm name of "Frey & Company," which said firm will collect all bills and pay all indebtedness.

Dated November 1st, 1897.

FRANZ A. FREY,
RICHARD LENZ.

A short-termed partnership.
Marin County Tocsin
November 2, 1897

FREY & COMPANY

The partnership of Frey & Lenz was short-lived. Richard Lenz had been a member of the Lang Brewing Company in San Francisco and appears to have returned to that firm. With all the short partnerships, Frey wanted to become strictly independent of Lang.

DECLARATION BY FREY & CO.

THE UNDERSIGNED, FRANZ FREY, UNDER the name of Frey & Company, at the City of San Rafael, County of Marin, State of Californie, is engaged in bottling and selling beer, lager bee and other beverages in bottles with his name, marks and devices branded, stamped, engraved etched. blown, impressed or otherwise produced upon such bottles, and

Under that certain act of the Legislature of California, entitled "An Act to Protect the Owners of bottles, boxes, siphons and kegs used in the sale of soda waters, Mineral or aerated waters, porter, ale, cider, ginger ale, milk, cream, small beer, lager beer, weiss beer, beer, white beer or otherbeverages," approved March 31st, 1891; the said Franz Frey makes the following descriptions and gives the following notices:

Notice is hereby given by said Franz Frey under and pursuant to said Act that in his said business he uses the following names and other marks blown or otherwise produced upon certain of the bottles in which his said beverages are bottled or sold, to wit:

Upon the surface of the bottle is blown or otherwise produced on the side thereof in a curved line the word "PROPERTY", beneath which in a straight line is the word "OF", beneath which in a straight line is the word, character and abbreviation "FREY & CO", beneath which in a straight line is the name of said City "SAN RAFAEL".

The following is a copy of said names or marks reduced in size, viz;

Marin County Tocsin
June 21, 1901

LISTING AND DECLARATION OF PROPERTY

PROPERTY OF FREY & Co. SAN RAFAEL

And notice is hereby further given by said Franz Frey under and pursuant to said Act that in his said business he uses the following additional names and other marks blown or otherwise produced upon certain other of the siphons in which his said beverages are bottled or sold, to wit:

Upon the surface of the bottle is blown or otherwise produced on the side thereof in a curved line the name, character and abbreviation "FREY & CO.", beneath which in a straight line is the name "SAN RAFAEL", beneath which in a straight line is the abbreviation "CAL."

The following is a copy of said names and marks reduced in size, viz:

FREY & CO. SAN RAFAEL CAL.

Dated, San Rafael, June 21, 1901.

FRANZ FREY.

Office of the County Clerk
of the County of Marin, } ss.
State of California.

I, ROBERT E. GRAHAM, County Clerk of the County of Marin and State aforesaid, and ex-officio Clerk of the Superior Court thereof, do hereby certify that I have compared the foregoing copy of Notices and Descriptions, under Statutes 1891, chapter 154, of Names, Marks and Devices on Bottles and Siphons by Franz Frey, and of the endorsements thereon, with the original record of the same remaining in this office, and that the same is a correct copy thereof, and of the whole of said original record.

WITNESS my hand and the seal of said Court this 21st day of June, 1901.

10 c. revenue cancelled  (SEAL)

ROBT. E. GRAHAM,
Clerk.
By F. S. HOLLAND,
Deputy Clerk.

FREY & COMPANY

Marin County Tocsin
Post partnership advertisement
July 1, 1899

**Frey & Co. bottlers of Wieland's Lager advertisements from 1910 Marin Directory.
Dr. Thomas Jacobs collection.**

PROPERTY
OF
FREY & Co.
SAN RAFAEL

Quart Beer　　　　　　**Pint Beer**　　　　　　**Half Pint Beer**

Althought all Frey & Co. San Rafael beer bottles are rare the half-pint bottle on the far right is one of two known. Half-pint courtesy of Dan Brown is extremely rare especially in mint condition with correct porcelain stopper.

**PROPERTY
OF
FREY & Co.
SAN RAFAEL**

Crown Top

Porcelain Stopper

S F & P G W
(San Francisco & Pacific Glass Works)

Franz Frey passed away April 6, 1922 cause of death being cancer of the throat.

FREDERICKSBURG BEER
FREY & CO. AGENTS

This one of a kind 1910 paper labeled Fredericksburg lager beer listing Frey & Co. as sole distributors is in the collection of Dr. Thomas Jacobs.

PRODUCTS SOLD BY FREY & CO.

Marin Journal
December 22, 1898

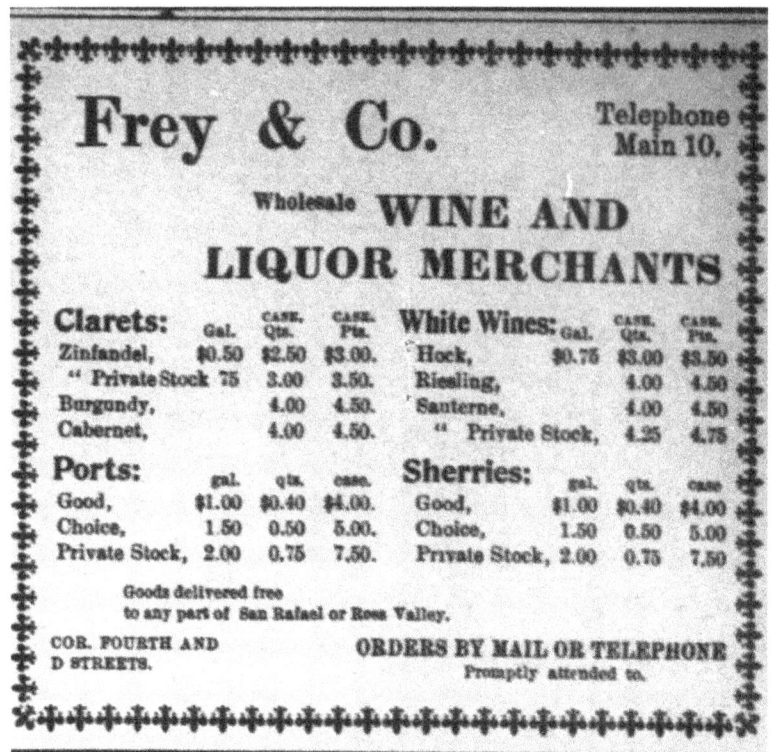

Marin Journal
February 23, 1899

> The dangers of civilization are overwork, worry and germs. We need a vitalizing power to sustain us. Learned men and experience point to pure whiskey. Primitive men did not need whiskey. We do. Changed conditions bring fresh needs And THE whiskey is HARPER.
> SOLD BY
> FREY & CO.
> San Rafael, Cal.

I. W. HARPER WHISKEY SOLD BY FREY & CO.

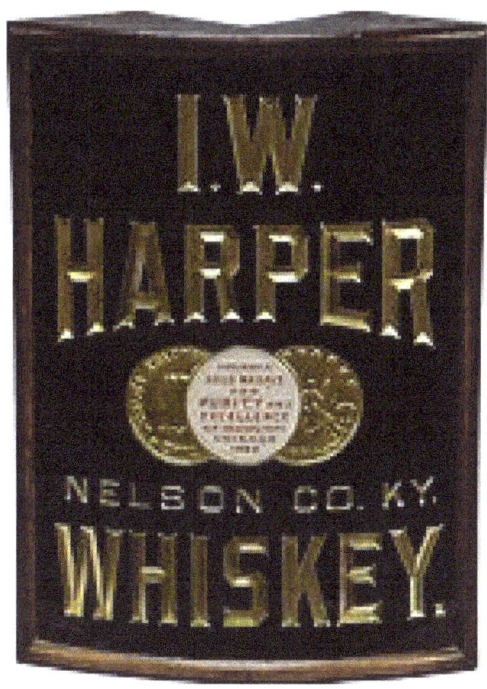

Corner Sign

Celebrated Harper Whiskey Label

FREY & CO. MERCHANDISE

WINES FOREIGN and DOMESTIC

Residents and summer guests will find the best wines of all grades in our cellars, the cheapest and the dearest. No need to go to San Francisco for quality or price in our line.

☞ *Also, we carry every class of goods Pertaining to the wholesale and retail liquor trade.*

FREY & CO.
Fourth and D Sts.,   San Rafael, Cal.

June 10, 1899
Marin County Tocsin

Strength-Giving

That's one unvarying characteristic of all our fine liquors; they feed a weakened body as pure air feeds the lungs. Delicious in flavor, reliable in quality — they make friends and keep them.

..FREY & CO..

Wholesale and Retail Liquor Dealers

Telephone Main 10   Fourth and D Sts.

August 10, 1901
Marin County Tocsin

TO YOUR HEALTH

December 21, 1901
Marin County Tocsin

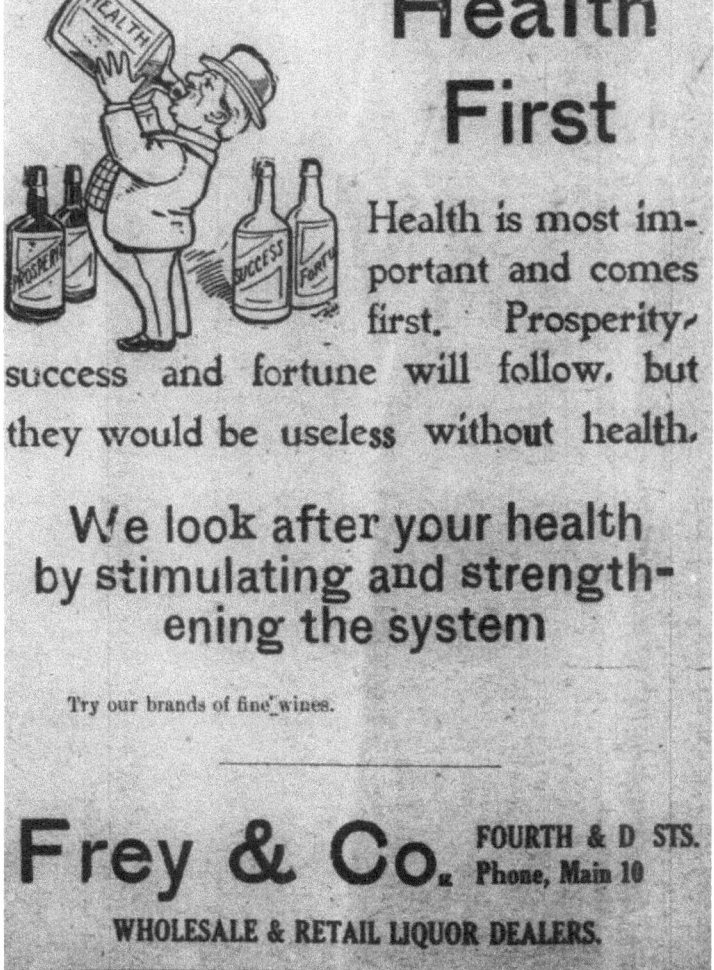

January 4, 1908
Marin County Tocsin

WINE ADVERTISEMENT

May 16, 1908
Marin County Tocsin

WINE ADVERTISEMENT

WINE FOR THE FAMILY TRADE

All our Wine Bought Direct from the Wineries of Napa and Sonoma Counties

Age and Purity Guaranteed

Give us a trial on any of our Red or White Wines and you will not be disappointed.

Prices are Moderate

A FEW SUGGESTIONS.—Zinfendels, Burgundies, Cobernets, Port or Sherries (domestic *and* imported), Rieslings and Sauternes.

Agents for JOHN WEILAND'S LAGER BEER. The Best Beer Produced in California.

Delivery Free to all Parts of Marin County

FREY & CO.

Wholesale and Retail Liquor Dealers

Phone 10   San Rafael   901-905 Fourth Street

January 4, 1913
Marin County Tocsin

OLD KENTUCKY RYE

February 28, 1908
Marin County Tocsin

Question:
Where did we get the expression "O.K."?
Answer:
From early liquor store dealers.
They were always asked if the whiskey was from "Old Kentucky" and shortened their answer to "It's O. K." meaning Old Kentucky.

FREDERICKSBURG BEER
FREY & COMPANY AGENTS

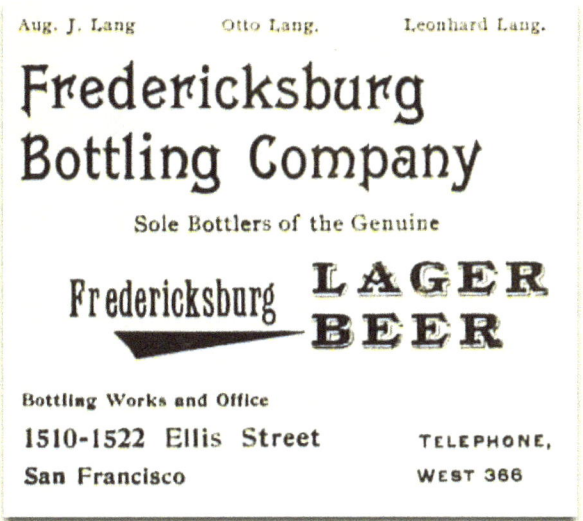

Bottled by Fredericksburg in San Francisco for Frey & Co.

ADDITIONAL PRODUCTS SOLD BY FREY

Among his merchandise was beer bottled by Fredericksburg Brewery of San Francisco, mineral waters by Aetna, Jackson, and Geyser. Franz Frey passed away April 6, 1922; cause of death being cancer of the throat.

AETNA MINERAL WATER

JACKSON NAPA SODA SPRING'S

GEYSER SODA WATER

GEYSER NATURAL BOILED MINERAL WATER

FRANETTA WHOLESALE LIQUORS

- 1884 – 1888   John Franetta
- 1888 – 1891   George Franetta & George Nowell
- 1891 – 1892   George Nowell, Otto Lang, Leonard Lang, August Lang & W. O. Cohoon
- 1892 – 1920   Otto Lang, Leonard Lang, August Lang & W. O. Cohoon

JOHN FRANETTA
LIQUOR & WINE MERCHANT

FRANETTA WINE & LIQUOR STORE
Photo courtesy of Newall Snyder collection

Close-up view of John Franetta with horse & buggy
Enlargement from Newall Snyder's photo

ADVERTISEMENTS

J. FRANETTA, wholesale dealer in wines and liquors, Fourth and D streets, San Rafael, keeps the best goods in his line, and will supply the trade to order, at rates below those of the city.

Marin Journal
September 20, 1883

MR. FRANETTA has just received forty casks of California Claret and Zinfandel from St. Helena. These are the very best of California wines, three years old; also 200 barrels Deer Lodge whiskey, for which he is sole agent on this coast. Franetta handles the best of liquors, and his customers can rely on the brands.

Marin Journal
May 1, 1884

ADVERTISEMENTS

Fredericksburg LAGER BEER AGENCY.

NOTICE IS GIVEN TO THE PEOPLE OF San Rafael and Marin County that the undersigned are the Sole Agents in this state for the justly celebrated

FREDERICKSBURG LAGER,

The Best and Most Wholesome Beer on the Pacific Coast, and that

MR. J. FRANETTA

Is our duly authorized and only Agent for this Beer in Marin County. Any other Beer sold for Fredericksburg is a fraud, and not the genuine and excellent Fredericksburg at all.
LANG BROTHERS,
San Francisco.

**Marin Journal
October 1885**

THE CELEBRATED Fredricksburg Beer

Is the most invigorating tonic, the most agreeable beverage, and the healthiest malt drink in the market.

Bottled by LANG BROS.,

The oldest and best bottlers on the Coast, and sold by

JOHN FRANETTA,

Cor. Fourth and B Sts., SAN RAFAEL,

SOLE AGENT FOR MARIN COUNTY.

**Marin Journal
August 9, 1888**

FRANETTA ADVERTISEMENT

WHOLESALE Wine
— AND —
Liquor HOUSE,

CORNER FOURTH AND D STS.,
SAN RAFAEL.

To the Trade and for Family Use. Of the Very Highest Quality, and at San Francisco Prices.

The trade of San Rafael and Marin County will positively find it to their advantage to buy of me. I ask a trial.

JOHN FRANETTA

JOHN FRANETTA WHISKEY BOTTLE

J. FRANETTA
WHOLESALE WINE &
LIQUOR DEALER
COR. 4TH & D ST.

Rick Siri Bottle

Advertisement from Dr. Thomas Jacobs collection

NOWELL & FRANETTA

John Franetta sold his liquor store to his son George Franetta and George Nowell. As listed in the ad below they continued to carry wines, Kentucky whiskies, Dublin Porter and Scotch Ales. Location remained at the corner of 4th & D streets. The partnership dissolved August 14, 1891.

Nowell & Franetta

Are importers and wholesale wine and liquor dealers. It has no sample room or bar. It is exclusively wholesale, and one of the most complete and most respectable houses in its line in the State. Messrs. Nowell & Franetta are also general commission merchants. The firm is composed of George Nowell and George Franetta. They are successors to Mr. J Franetta, who established the house eight years ago. They are importers and carry a fine stock of pure wines and old Kentucky whiskies, Dublin porter and Scotch ales. They furnish the sideboards of the wealthy families who reside here, and can undersell the jobbers of San Francisco. The establishment is large, well stocked, and has a prosperous appearance. Located at corner Fourth and D streets.— Journal of Commerce.

NOTICE OF DISSOLUTION OF CO-PARTNERSHIP
George Nowell & George Franetta

The firm of Nowell & Franetta has dissolved, George Franetta (John Franetta's son) retiring and his interest purchased by Lang Bros., well known business men of San Francisco. The firm is reorganized under the title Nowell, Lang Bros. and Company and will continue business as formerly, at the old stand on the corner of Fourth and D Streets San Rafael.

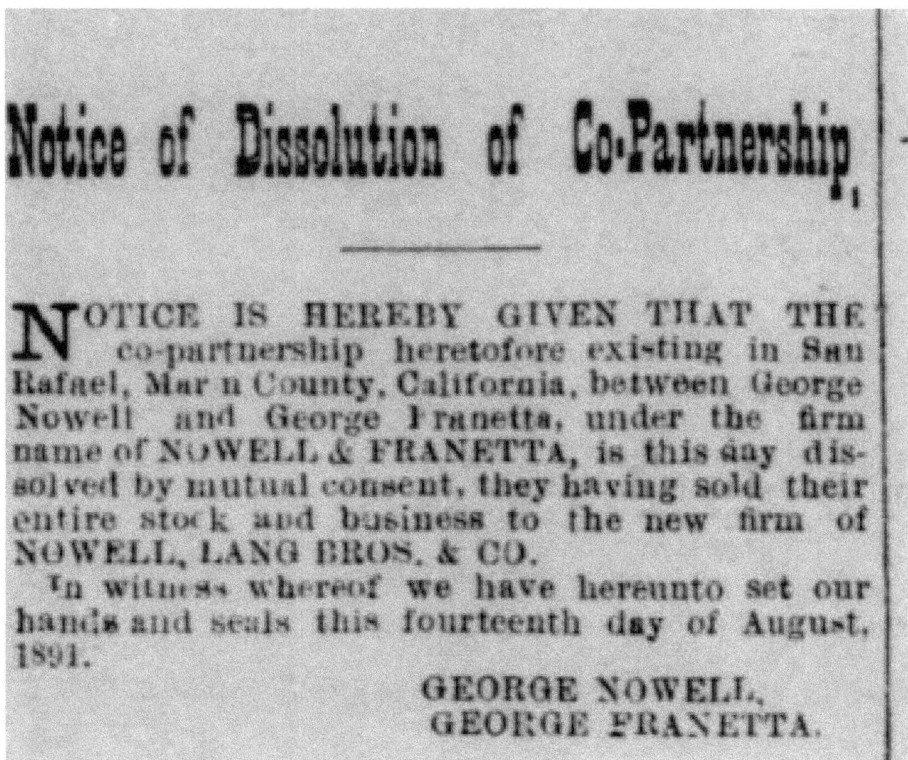

Marin County Tocsin
August 15, 1891

CO-PARTNERSHIP
Lang Bros., Cohoon, and Nowell

With the ending of the partnership of Nowell and Franetta a new partnership was formed the same day; August 14, 1891, between George Nowell, W. O. Cohoon, and the Lang Brothers, Otto, Leonhard, and August.

NOTICE OF CO-PARTNERSHIP.

WE THE UNDERSIGNED DO HEREBY certify that we are co-partners doing business in the town of San Rafael, in the county of Marin, State of California, under the firm name of NOWELL AND LANG BROS. & Co., and that the names in full of all the members of said partnership are as follows:

W. O. COHOON,
Residing at the Town of San Rafael in Marin County, State of California.

OTTO LANG,
By August Lang his Attorney in fact, Residing in the City and County of San Francisco, State of California.

LEONHARD LANG,
By August Lang his Attorney in fact, Residing in the City and County of San Francisco, State of California.

AUGUST LANG,
Residing in the City and County of San Francisco, State of California.

GEORGE NOWELL,
Residing at the town of San Rafael, in the County of Marin, State of California.

In witness whereof we have hereunto set our hands this 14th day of August, 1891.

OTTO LANG,
LEONHARD LANG,
AUGUST LANG,
W. O. COHOON,
GEORGE NOWELL,

Marin County Tocsin
August 15, 1891

STATE OF CALIFORNIA, } ss.
COUNTY OF MARIN,

On this 14th day of August in the year one thousand eight hundred and ninety-one, before me, R. P. Troy, a Notary Public in and for said county of Marin, residing therein, duly commissioned and sworn, personally appeared W. O. Cohoon, August Lang and George Nowell, known to me to be the same persons described in, whose names are subscribed to, and who executed the within instrument and they acknowledged to me that they executed the same.

In witness whereof, I have hereunto set my hand and affixed my official seal at my office in the County of Marin the day and year in this certificate first above written.

[SEAL]
      R. P. TROY,
      Notary Public.

STATE OF CALIFORNIA, } ss.
County of Marin.

On this 14th day of August, in the year one thousand eight hundred ninety-one, before me, R. P. Troy, a Notary Public, in and for said county, residing therein, duly commissioned and sworn, personally appeared August Lang, known to me to be the person described in, and whose name is subscribed to the within instrument, as the attorney in fact of Otto Lang and Leonhard Lang, and the said August Lang acknowledged to me that he subscribed the names of Otto Lang and Leonhard Lang thereunto as principals and his own name as attorney in fact.

In witness whereof I have hereunto set my hand and affixed my official seal at my office in the said county of Marin the day and year in this certificate first above written.

[SEAL]
      R. P. TROY,
      Notary Public.

Endorsed: Filed in the office of the County Clerk of Marin county, State of California, this 14th day of August, 1891.
    THOS. S. BONNEAU.
      Clerk.

DISSOLUTION OF PARTNERSHIP

The former partnership of Lang Bros., Cohoon, and Nowell lasted less than a year when George Nowell retired leaving the Lang Brothers and W. O. Cohoon as proprietors.

Dissolution of Partnership.

THIS IS TO CERTIFY THAT THE FIRM OF Nowell, Lang Bros. & Co., doing business in the town of San Rafael, county of Marin, has dissolved by mutual agreement, George Nowell retiring, and that Messrs. Lang Bros., and W. O. Cohoon, will carry on the wholesale wine and liquor and beer-bottling business, under the firm name of Lang Bros. & Co., and pay all debts and collect all outstanding accounts as the books show of the late firm.

LANG BROS.,
By OTTO LANG.
GEO. NOWELL,
W. O. CAHOON.

San Rafael, June 28, 1892.

DISSOLUTION OF PARTNERSHIP

Dissolution of Partnership.

C. O. CAHOON, of the firm of Lang Brothers & Company, having sold out his interest in said firm to Messrs. Lang Brothers, that firm will continue business at the old stand on Fourth street, San Rafael Cal., under the name of Lang Bros and Co., and will pay all debts and collect all outstanding bills due the firm when Mr. Cahoon was a partner.

Signed this 12th day, Feb. 1894.

LANG BROS. &CO.

a-b1sf55.1

With the resignation of W. O. Cahoon the business now is under full control of August J. Lang and his brothers. They operated the business in San Rafael from February 12, 1894 until they sold to Franz Frey and Richard Lenz in January 1897.

LANG BROS. & CO.

**1894 Lang Bros. billing the Tamalpias Academy for $1.60
Newall Snyder collection.**

Lang Bros. operated the Philadelphia Bottling Company and had depots in San Francisco, San Rafael and possibly Napa.

**August J. Lang & Co. San Francisco beer bottle.
Face of two Philadelphia Bottling Co. bottles in Dan Brown collection.**

LANG & CO.

- August Lang is listed in American Breweries book as being in Napa in 1880, however, it is believed that he applied but never opened a brewery or bottling plant there. It probably was a depot for Philadelphia beer.

- Brother Adolph was brewing beer in Pendleton, Oregon from 1878 until 1884 as Adolph & Company.

- In 1880, Adolph & Otto Lang opened business at 1406 Polk Street bottling John Wieland's Philadelphia beer. August joined Adolph & Otto in 1887 bottling beer. Adolph left the firm and went independent bottling beer.

- 1890, the brothers formed the Fredericksburg Bottling Company at 1510-1512 Ellis Street with Otto president; Adolph, who had been a brewer in Pendleton, Oregon, vice-president; Leonard, who had been a baker foreman; and August, both manager's.

- In 1892, Wilhelm, was manager of the Lang Brother's Oakland branch. He left the family business in 1898 and managed the Oakland Pioneer Soda Water Company.

- 1894 Lang Bros.; purchased the Frey Bottling Company in San Rafael.

- 1899, Adolf left the Lang Brothers and started the National Bottling Company on Polk Street, relocating back to 1510-1512 Ellis Street until the San Francisco Earthquake and Fire.

- After the "Quake & Fire" August bought out all brothers and formed the August Lang & Co., which bottled Fredericksburg beer, moving to 18th and Alabama Streets.

- Moving to San Rafael in 1902, August commuted to San Francisco. September 1911, August took control of the Red Lion Brewery, corner of Geary & Baker Streets. Here they produced Red Lion Ale and Porter.

- August sold his Fredericksburg Bottling Company to San Francisco Breweries, Inc. in 1912 and started a real estate company.

MARIN COUNTY CLUB
WINE & LIQUORS

- 1905 – 1915   Col. Stubbs

> Col. Stubbs, the genial proprietor of the Marin Club, remembered each of the club members on Xmas Day. The Club, under the Colonel's management, is growing in popularity as it is a desirable place for the young men of the city to meet and exchange ideas.

December 29, 1906
Marin County Tocsin

> Col. Stubbs has started a new enterprise. He has rented the new Corti building on Fourth street and opened the Unique Coffee and Lunch parlor. It is a neat and clean place, and coffee and light lunches alone are served.

July 22, 1909
Marin Journal

Col. Stubbs of Larkspur has leased the club rooms in the Wilkins block and installed billiard and pool tables in the gymnasium. He has organized a big club.

October 25, 1906
Marin Journal

S. A. Pacheco

Telephone Black 711

Wholesale and Retail

Wine and Liquor Merchant ..

Agent for Rainier Beer and Marin County Club Whiskey

703 Fourth Street   San Rafael, Cal.

October 20, 1901
Marin County Tocsin

MARIN COUNTY CLUB BOTTLES

**ELL-ELL bottle American Bottle Auctions Sacramento
Clear bottles on each end John Louder collection**

SALVADORE GRANDI
Point Reyes Station

At a recent Northwestern Bottle Collectors meeting we had "Show & Tell" that included whiskey flasks. Eric McGuire brought in a fabulous paper label bottle from Salvadore Grandi, a red brick building downtown on Main Street Point Reyes. Built in 1883 by A. P. Whitney of Petaluma Grandi purchased the building in 1887.

Olema Station was originally founded in 1875 being a train station and renamed Point Reyes on August 10, 1891. The narrow gauge railroad hauled lumber from Cazadero to the ferry in Sausalito.

The building housed a general store, post office, dance hall, hotel, restaurant and saloon. In 1906 the San Francisco Earthquake leveled Point Reyes including Grandi's red brick building.

Grandi rebuilt the building and operated a hardware store, hotel, food and beverage operation. It must have included an off-sale liquor license as we see by this great paper label bottle.

The railroad shutdown service in 1950 taking most of Point Reyes commerce with it. The building was closed and vacated in 1970.

Point Reyes National Seashore Archives

GRANDI'S SPECIAL WHISKEY
BOTTLED BY THE GRANDI COMPANY
Point Reyes Station, Calif.

Eric McGuire collection

JACK BRADY
SEQUOIA, MILL VALLEY

JACK BRADY OPENS MILL VALLEY SALOON

Mill Valley ceased to be a "dry town" last Sunday. In the morning Jack Brady opened his saloon and throughout the day did a rushing business. He pays $750 a year for his license and therefor operates the only saloon in the town. He got his license some two months ago, but was unable to secure a building in which to open. There was a lawsuit over a lease, and a lot of other legal trouble before he could open.

Mr. Brady was one of the most popular conductors in the employ of the North Shore railroad.

June 29, 1905

J. E. BRADY PUMPKINSEED

J. E. BRADY
SEQUOIA
MILL VALLEY, CAL.

J. E. BRADY
SEQUOIA
MILL VALLEY, CAL.
NET CONTENTS 4 OZ.

The Brady Pumpkinseed comes in both 4 oz. and 8 oz. sizes and both with and without contents on bottle face. The 8 oz. size comes with a double ring on top.

These two pumpkinseeds compliments of Dan Brown collection.

JACK E. BRADY FIFTH

**J. E. BRADY
SEQUOIA
MILL VALLEY, CAL.**

HENRY C. GIESKE
FAMILY
LIQUOR STORE
300 B ST. COR. 2nd
SAN RAFAEL, CAL.

Dandy flask compliments of Chuck Ingraham collection.

THE SENATE
S. KERRIGAN
PROP.
SAN RAFAEL, CAL.

Pumpkinseed compliments of Dan Brown collection.

A. TIMONY

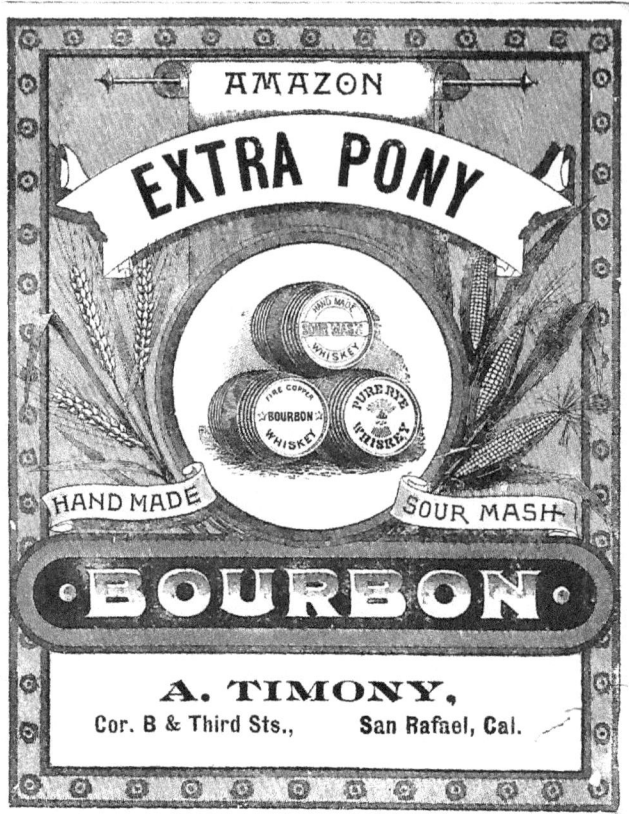

A. Timony Reopens His Place at Los Gallinas

A. Timony has reopened his road house at Los Gallinas and for awhile will give the business his personal supervision. It will remind one of old times to again see the popular and veteran entertainer at the old stand.

TURKEY SHOOT AT TIMONY RANCH

CANDIDATES' Shooting Match!
AT A. TIMONY'S,
SANTA MARGARITA,
POSTPONED TO
SUNDAY, OCT. 26, '84,

TURKEYS, CHICKENS, AND PIGEONS,
Glass Balls, Etc.
THE PUBLIC ARE INVITED.
A. TIMONY, Proprietor.

October 16, 1884
Marin Journal

A. Timony has sold his saloon at Third and B streets to Joseph Lafranchi.

Joseph Lafranchi purchased the Dewey Saloon on the corners of Third & B Streets September 1912 from Timony.

September 26, 1912
Marin Journal

EAGLE BAR
Sausalito, Cal.

**Compliments of ..
EDDIE GOLDEN
EAGLE BAR**
Sausalito, Cal.

Half pint, tooled screw top Dandy flask covered with paper mâché. This flask was a gift to special customers on birthdays and holidays.

John Louder collection

GOLDEN EAGLE HOTEL
SAN RAFAEL

**Notice the San Rafael Steam Beer corner sign.
Photograph Rick Siri collection.**

GOLDEN EAGLE HOTEL
SAN RAFAEL

Half pint, tooled screw top Dandy flask covered with paper mâché.

Newall Snyder collection.

ZOPF'S WINE GARDENS

Rare trade card for Hermann Zopf Wine Gardens in San Rafael.
Newall Snyder collection.

Marin Journal Advertisements

July 1876

BOCA BEER,
AT THE WINE GARDENS OF
HERMANN ZOPF.

May 1878

ZOPF'S WINE GARDENS,
Near H Street Station,
SAN RAFAEL.
EVERY SUNDAY AFTERNOON AND
EVENING
ZITTA CONCERT.

BEAUTIFUL FLOWER GARDENS, WITH RUStic arbors and retreats, a delightful resort.
HERMANN ZOPF.

CHAPTER 2
MARIN COUNTY SODA WORKS

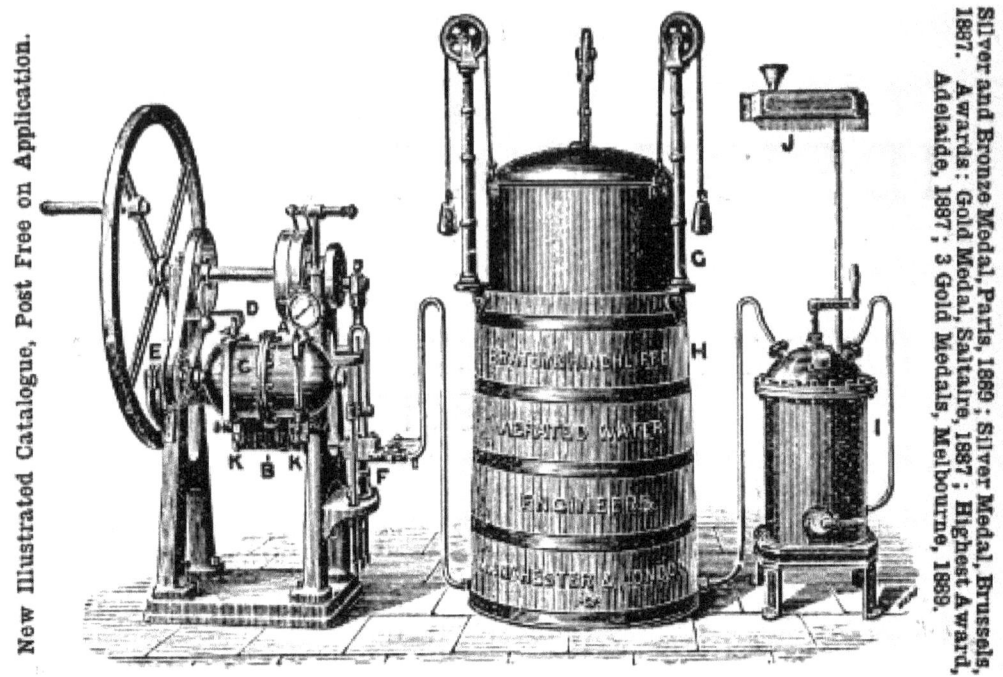

EXAMPLES OF SODA MANUFACTURE MACHINES

SAN RAFAEL SODA WORKS
Corner 4th & H Street

- 1879 – 1883 Joseph Kappenman

Joseph Kappenman was a saloonkeeper in San Rafael who also bottled soda water.

FACE

**SAN RAFAEL
SODA WORKS
J. KAPPENMAN
PROP'T**

REVERSE

BLANK

Bottle image courtesy of American Bottle Auctions

CENTRAL HOTEL SALOON
Fourth & C Streets

**Frank Jacob tending bar at Kappenman's saloon in the late 1880's.
Joseph Kappenman facing bar.
San Rafael Library Collection**

SAN RAFAEL SODA WORKS

ADVERTISEMENT
Marin Journal, Volume 19, Number 14, June 12, 1879

May 3, 1883 Kappenman purchased interest in a lot on the southeast corner of 4th & C Streets San Rafael from W. A. Bray for $3,100.

The following liquor licenses were granted: Joseph Kappenman, H. Ferrario, and R. F. Wilde of the Central Hotel October 6, 1906.

Joseph Kappenman has installed a new modern bar in his place and will soon remodel his building, putting in a new front and making other improvements September 26, 1907

SAN RAFAEL SODA WORKS

- 1884 – 1886   Sylvan Provencal & Alphonse Bresson

Roommates Sylvan Provencal & Alphonse Bresson purchased the San Rafael Soda Works from saloonkeeper Joseph Kappenman. P & B embossed on the bottle relates to Provencal & Bresson.

I emailed Tom Jacobs about this bottle as I thought it may be Peterson & Borello. He said he thought so for years also, however, there was a Samuel's Napa Soda bottle brought this attention that has A & B on the bottom.

Tom did research and the 1880 Marin census shows A. Bresson as Napa Soda agent who lived in Fairfax. Further checking revealed that Bresson was associated with a man named Provencal in the soda business in San Rafael in 1880's.

So this bottle must be attributed to Provencal & Joseph Bresson. In our mineral water book *"LAKE, NAPA, SONOMA, MENDOCINO, SOLANO, MARIN, and HUMBOLDT MINERAL & HOT SPRINGS"* the Siri bottle A & B is the Napa soda attributed to Bresson who was the Marin county agent for Samuel's.

**SAN RAFAEL
SODA WORKS
P & B
PROPS's**

Bottle image courtesy of American Bottle Auctions

MARIN SODA & IRON PHOSPHATE WORKS

- 1886 - 1901   Martin Petersen
- 1901 – 1905   Eugene Klammer & Eugene Malz
- 1905 – 1911   Eugene Malz

Martin Petersen, an immigrant from Schleswig, Germany, opened a bakery in San Rafael in 1872. He then married Mary Kelly in 1877, and fathered six children.

In 1886 his soda works were located on the corner of First & D streets as was his home. Not only did he manufacture his own soda, he was an agent for Jackson's Napa Soda, Geyser Soda, and Aetna Soda Water.

Petersen's water was available in many flavors; lemon, lime, orange phosphates, fruit flavored champagne and his specialty named Horehound, a combination of Honey & Lime Juice. By 1888 he was wholesaling his products throughout the state of California. He also opened an office at 415 Battery Street in San Francisco run by Fred Schenkel.

His Iron Phosphate soda was his main product and best seller. It was pure bicarbonate instead of marble dust. Early methods of creating carbonated water was to combine marble dust with sulphuric acid passed into water creating the bubbles.

March 7, 1889, he appointed William W. Skaggs as an agent.

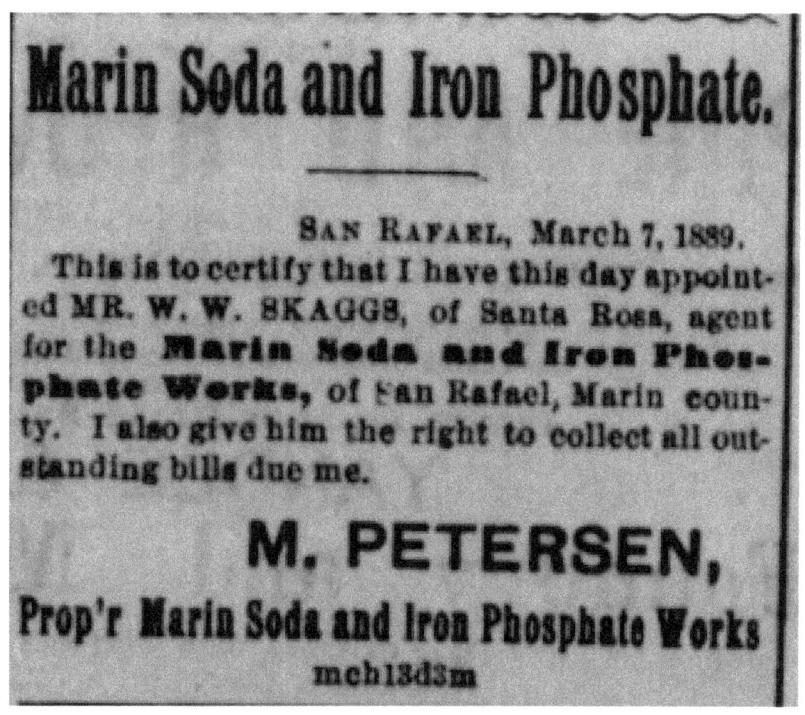

ADVERTISEMENT

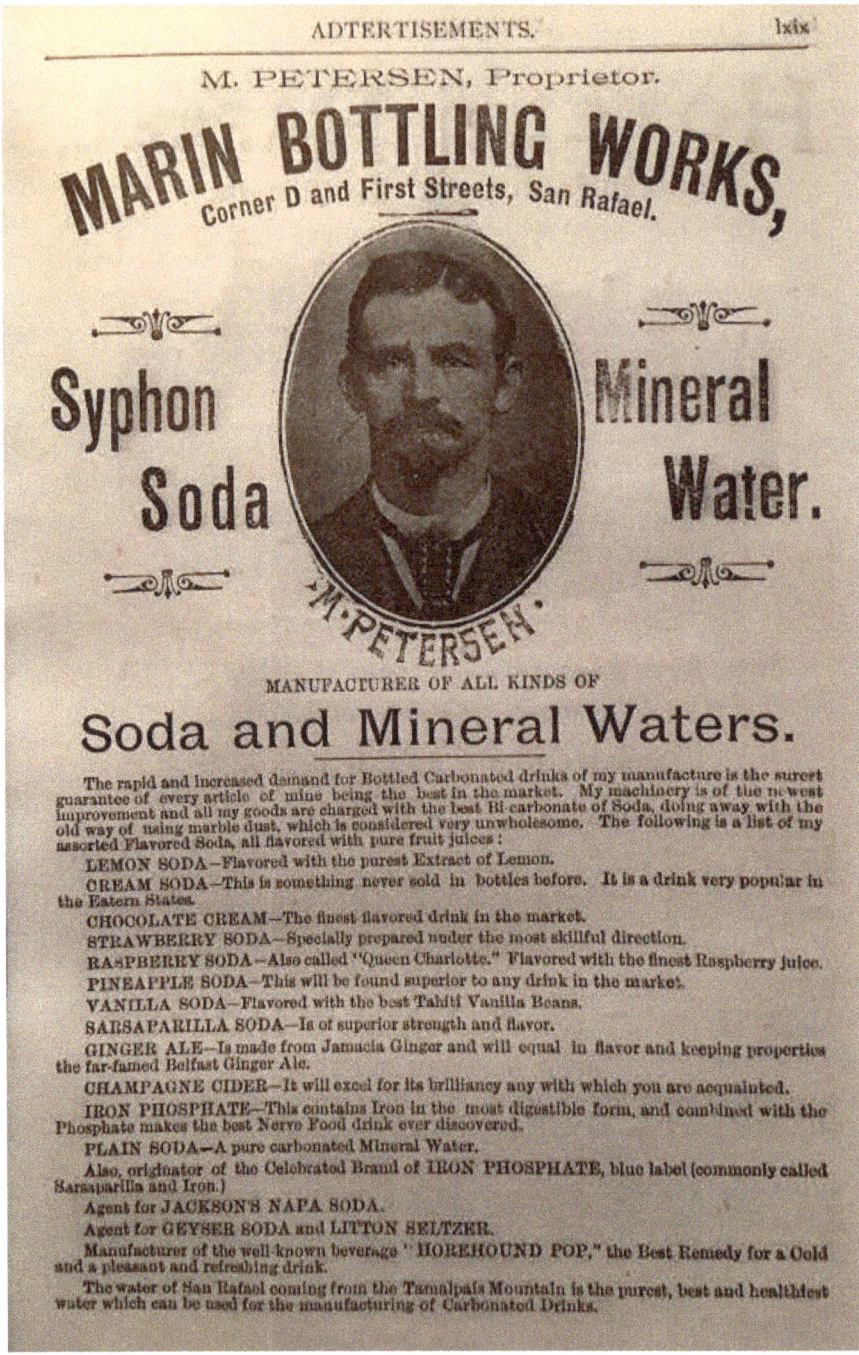

Advertisement from an extremely rare 1890 Marin Directory listing flavors of soda's using Tamalpais Mountain Water in addition to Jackson's Napa Soda, Geyser Soda and Litton (Lytton) seltzer waters carried by Peteresen.

Dr. Thomas Jacobs collection.

M. PETERSEN ADVERTISEMENT IN GERMAN

Newall Snyder collection.

MARTIN PETERSEN BOTTLE

**M. PETERSEN
MARIN
SODA WORKS
SAN RAFAEL**

On some of the Petersen Hutchinson style sodas the bottom is blank and some are embossed with MSW which stands for Marin Soda Works.

The bottle image above also comes misspelled: SAN RAFAL.

MARTIN PETERSEN BOTTLES

**MARTIN PETERSEN
SAN RAFAEL
CAL.**

**MARIN BOTTLING WORKS
MBW
(Logo)
S.R.CAL.**

NOTICE OF MANUFACTURING

NOTICE!

THE UNDERSIGNED BEGS LEAVE TO STATE that he is engaged in manufacturing, bottling and selling

Soda Waters, Mineral Waters, Ginger Ale

and other beverages, in bottles, siphons, kegs and boxes, in the town of San Rafael, Marin county, State of California, with his principal place of business situated in the town of San Rafael, Marin county, State of California, and the following is a description of the names and marks branded, stamped and blown upon the bottles, siphons, kegs and boxes used by him in said business as aforesaid, to wit:

Soda bottles marked on side of bottle "M. Petersen, Marin Soda Works, San Rafael;" soda bottles also marked "M. B. W." on bottom of bottle; soda boxes marked "Marin Soda Works, M. Petersen, proprietor, corner D and First streets, San Rafael;" siphons marked "Marin Soda Works," also marked "Marin Bottling Works;" boxes marked "Marin Soda Works, M. Petersen, proprietor, corner D and First streets, San Rafael;" copper kegs and tanks marked "M. S. W."

STATE OF CALIFORNIA, } ss
County of Marin }

On this 20th day of November in the year A. D. 1891, before me, James W. Cochrane, a Notary Public in and for the said county of Marin, personally appeared Martin Petersen, known to me to be the person whose name is subscribed to the within instrument, and he acknowledged to me that he executed the same.

In witness whereof I have hereunto set my hand and affixed my official seal the day and year in this certificate first above written.

(SEAL)   JAMES W. COCHRANE,
Notary Public.

Tocsin Notice
Dec 5, 1891

TRADEMARKS

The last Legislature passed a law to protect the manufacturers of soda and other mineral waters. It provides that such manufacturers may file with the County Clerk of their respective counties and with the Secretary of State a description of the various marks or devices on their bottles and packages and publish the same in a newspaper printed in county where the works are situated. Thereafter any other person who refills such bottles, etc., is guilty of a misdemeanor and becomes liable to a heavy fine. Martin Petersen, of the Marin Soda Works, has suffered considerably by parties selling imitation goods in his bottles and he therefore avails himself of this law. He has filed his descriptions with the County Clerk and Secretary of State and the necessary advertisement appears in this week's TOCSIN. Hereafter, he will make it interesting for parties who meddle with his business.

November 14, 1891
Marin County Tocsin

MARIN COUNTY BOTTLING WORKS

Marin County Bottling Works.

Martin Petersen, the energetic proprietor of the Marin County Bottling Works, is preparing to add a fine laboratory to his already extensive plant, an improvement which his rapidly extending business makes necessary.

The history of these works proves that San Rafael is as good as any other place to establish a manufacturing industry in, provided, of course, that the right man is at the helm. They started from a very humble beginning, but by energy and enterprise have in a very few years grown to be a big concern. As a manufacturer of the various sodas and fancy carbonated drinks, the proprietor, Mr. Petersen, has a foremost reputation in this State. His goods are standard and the big wholesale houses are only too glad to handle them. Mr. Petersen has not been satisfied with the narrow limits of this State. He has advertised his compounds thoroughly and is in receipt of big orders from Oregon, Nevada, Arizona, and eastward to the Mississippi. He even has a considerable line of foreign customers. The proprietor is thoroughly original. He devises himself the many palatable drinks that bear his trademark and is constantly inventing something new. His latest is a "champagne" that is by no means a poor substitute for the genuine article. It has all the delicacy of taste and pungency of true champagne, and you can drink a whole barrel full without wanting to steal a horse or engage in a free fight. It bids fair to be the greatest hit that Mr. Petersen has made during his successful career.

July 11, 1891
Marin County Tocsin

MARIN SODA & BOTTLING WORKS
Klammer and Malz

In 1895, William Petersen sold the Marin Soda and Bottling Works to two German immigrants, Eugene R. Klammer and Emil Malz. They had a soda works in Petaluma and also operated a soda business in San Rafael. They separated the business partnership in June 1906 with Malz continuing in San Rafael and Klammer the Petaluma operation.

HUTCHINSON STYLE BOTTLE

FACE	REVERSE
KLAMMER & MALZ SAN RAFAEL CAL.	Blank

CROWN TOP BOTTLES

KLAMMER & MALZ
TRADE **PET** MARK
PETALUMA
AND
SAN RAFAEL, CAL.
(Dan Brown bottle)

KLAMMER & MALZ
TRADE **PET** MARK
PETALUMA
CAL.

PET
ON BASE

MARIN
BOTTLING WORKS
E. MALZ
PROP.
SAN RAFAEL
CAL.

MBW
ON BASE

KLAMMER & MALZ SELTZER BOTTLE
Seltzer water was a large part of a soda manufacturers business.

KLAMMER & MALZ
SAN RAFAEL
CAL.

PUBLIC NOTICE

The firm of Klammer & Malz of the Petaluma and San Rafael bottling works has been dissolved. Mr. Malz will in the future conduct the San Rafael Establishment and Mr. Klammer will continue at the Petaluma branch.

TRANSFER OF TRADE MARKS

William Petersen sold all rights of Marin Bottling Works to Klammer & Malz in 1895. In August 1901 they sold all rights to Marin Bottling Works to the Borello Brothers of San Rafael.

August 17, 1901

THE UNDERSIGNED BROS.
I, HAVE TO STATE THAT HE IS ENGAGED IN MANUFACTURING, BOTTLING AND SELLING

Soda Waters, Mineral Waters, Ginger Ale

and other beverages, in bottles, siphons, kegs and boxes, in the town of San Rafael, Marin County, State of California, with his principal place of business situated in the town of San Rafael, Marin County, State of California, and the following is a description of names and marks branded, stamped and blown upon the bottles, siphons, kegs, and boxes used by him in said business as afore said, to wit:

Soda bottles marked on side of bottle "M. Petersen, Marin Soda Works, San Rafael;" Marin Soda Works, San Rafael; "Soda Bottles also marked "M. B. W." on bottom of bottle; soda boxes marked "Marin Soda Works, M. Petersen, proprietor, corner D and First streets, San Rafael; and copper kegs and tanks marked "M. Petersen."

Borello Bros.

BUFFALO BOTTLING WORKS
SAN RAFAEL

- 1890 - 1901   John Phelps
- 1901 - 1912   John, Frank & Andrew Borello

Marin County Library photo

Buffalo Bottling Works located on First and Hayes streets started approximately 1890 owned and operated by John Phelps and burned down September 22, 1896. He rebuilt the soda works and operated until March 31, 1901 when purchased by three brothers; Frank, John and Andrew Borello.

Frank Borello was part owner with Grant J. Porter of the Fresno Soda Works which was established in Madera, California a distance of about 25 miles. After Frank moved to San Rafael to work with his brothers, he relocated back to Fresno approximately 1907 and managed a winery.

Since there was a spring on the San Rafael property the Borello Bros. immediately dug a well and started bottling water. In 1904 they expanded by purchasing the neighboring property that was the Pioneer Mill & Lumber Company.

In 1911 the California Board of Health analyzed the "Famous Tamalpais Water" promoted by the brothers and found that their claim of the water being free of all foreign ingredients including organic or non-organic matter was false. The California Board of Health found sodium benzoate in their products which is a preservative in their bottled waters. The March 11, 1911 Sausalito News reported that warrants were issued for both John and Andrew Borello for violation of the 1906 Pure Food Law.

Mt. TAMALPAIS WATER ANALYSIS
Marin County Tocsin Advertisement – December 4, 1904

THE TAMALPAIS NATURAL MINERAL WATER SPRINGS, SAN RAFAEL, CAL.

The famous Tamalpais Natural Mineral Springs are located on BORELLO BROS.' property, a fac simile analysis of which herewith follows:

SAN FRANCISCO, December 17, 1904.

Messrs. Borello Bros., San Rafael, Cal.

Dear Sirs: We have made a careful analysis of the sample of water submitted to us by you, and find it to contain a Total Mineral Residue of 31.85 grains per gallon. Of this residue, 11.66 grains are Sodium Chloride, the remaining 20.19 grains consisting of Magnesium and Calcium Carbonates, which makes up the principal part, and a little Magnesium Sulphate. The water is absolutely free from all deleterious ingredients of either an organic or inorganic nature, and is well adapted for bottling and charging with carbonic acid gas. On account of its freedom from organic substances the water will keep well on bottling.

Very truly yours, THOMAS PRICE & SON,
Assayers and Analytic Chemists.

TO THE PUBLIC: The Tamalpais Natural Mineral Water is put up carbonated in syphons, large, medium and split bottles.

The Tamalpais Natural Mineral Water, uncarbonated, is put up in large bottles only.

Our carbonated beverages like Ginger Ale, Ironbrew, Lemon, Soda, Sarsaparilla, Hires', the genuine Root Beer, and other leading drinks, are put up in large, medium and split bottles, at reasonable prices, and are on sale at all groceries.

Residences and offices in San Rafael and vicinity supplied direct from our delivery wagons or through grocers. Country orders by railroad. Respectfully yours, **BORRELLO BROS.**

Marin County Tocsin Advertisement – October 24, 1908
FINE LITHOGRAPHIC VIEW OF SAN RAFAEL AND Mt. TAMALPAIS FROM BOYD MEMORIAL PARK

Borello Bros. proprietors of Tamalpias Natural Mineral Water have issued one of the handsomest calendars ever issued in California. It is a beautiful picture of San Rafael and Mt. Tamalpias as seen from the upper portion of Boyd memorial Park.

It is natural colors showing portions of Boyd Park, Mr. Boyd's residence, many of the city's prominent buildings, including school houses and churches, the forest covered hills which skirt the southern portion of our city, and the Tamalpias Mountain Peak in the distance.

It is worthy of a place on any wall. Borello Bros. deserve praise for getting out so credible an advertisement of San Rafael and of their mineral water.

DIRECTORY BACK COVER

Telephone Black 334　　　　　　　　　　Postoffice Box No. 221

BUFFALO
..SODA WORKS..

BORELLO BROS.
Props.

Lemon and Cream Soda, Sarsaparilla, Sarsaparilla and Iron, Ginger Ale
Champagne Cider, and Syrup of All Kinds

Sole Agents for the Celebrated Iron Brew

Soda Fountain Tanks Charged — Syphon Soda and Mineral Waters

.....AGENTS FOR.....

California Natural Carbonic Acid Gas Co.

Agency for Samuel Soda Springs. These Springs are located in Napa County, 20 miles east of St. Helena. The Water is bottled at the Springs and contains its own Natural Gas.

Our Ginger Ale, Sarsaparilla and Root Beer
are equal to any imported.

-------- PRICES ON APPLICATION --------

OFFICE AND WORKS
1119 FOURTH ST.　　San Rafael

TAMALPAIS NATURAL MINERAL WATER

FACE

MT. TAMALPIAS
NATURAL
MINERAL
WATER CO.
SAN RAFAEL CAL.

REVERSE

A MILD
MINERAL WATER
T	TAMALPIAS	M
R	SODA	A
A	FREE FROM ALL	R
D	DELETERIOUS	K
E	INGREDIENTS	

BORELLO ADVERTISEMENTS

Marin County Tocsin Advertisement
September 3, 1907

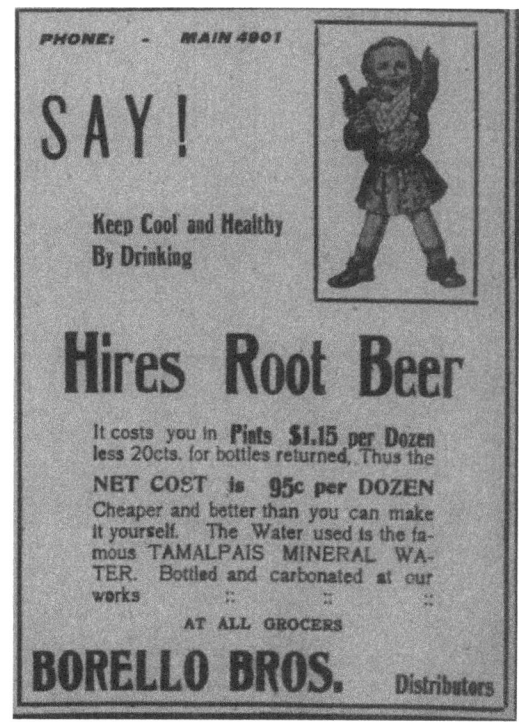

Marin County Tocsin Advertisement
September 23, 1911

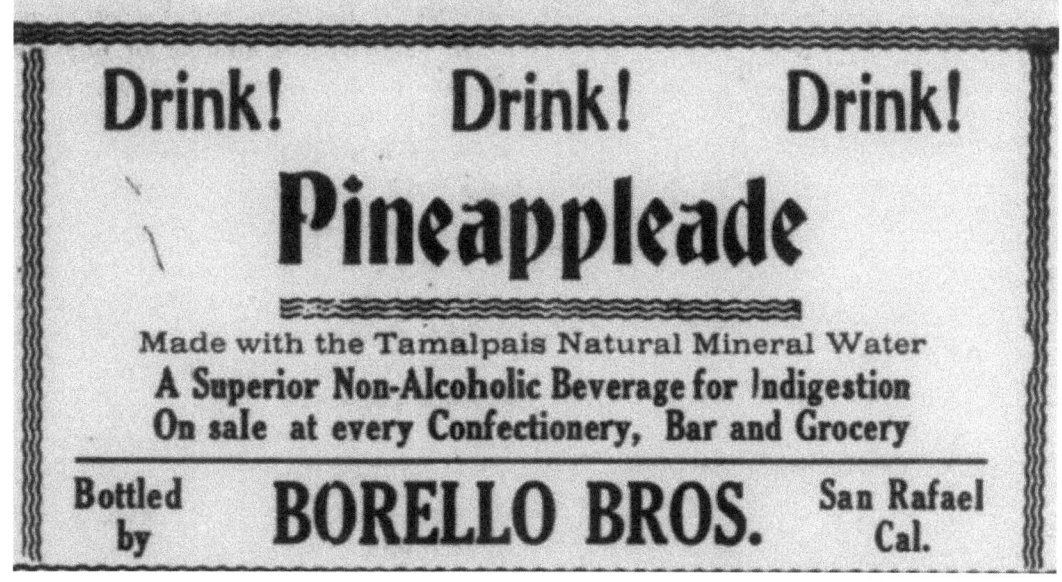

BORELLO ADVERTISEMENTS
Marin County Tocsin Advertisement
June 3, 1911

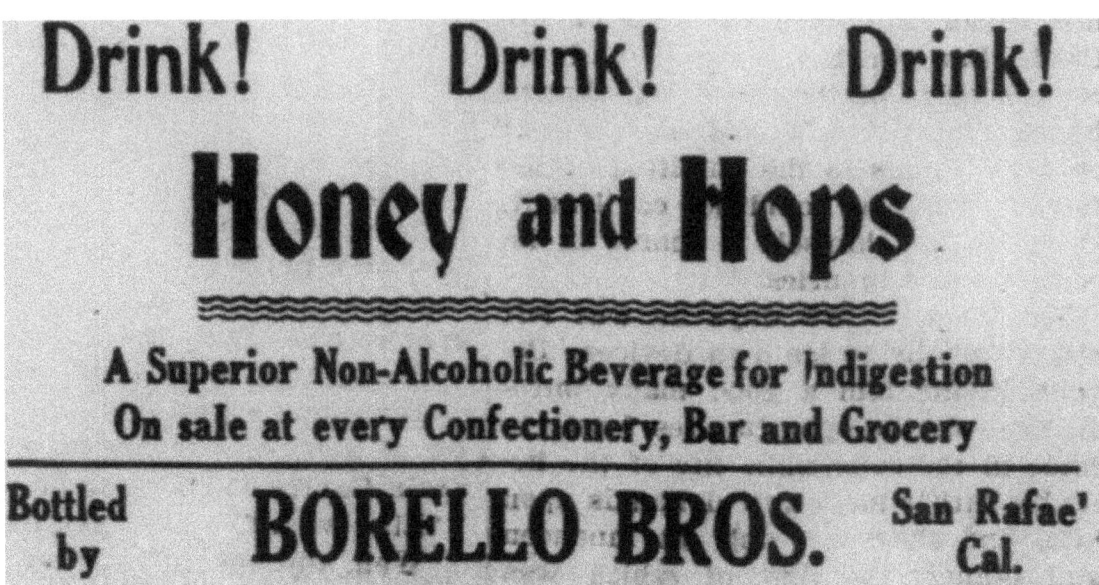

Marin County Tocsin Advertisement
July 1, 1899

BORELLO BROS.
HUTCHINSON BOTTLES

FACE

BUFFALO
BOTTLING
WORKS
B. B. W.
SAN RAFAEL
CALA.

REVERSE

BLANK

BOTTOM
Some blank and some
with 4 Point Star

FACE

B. B. W.
SAN RAFAEL

REVERSE

BLANK

BORELLO BROS.
CROWN TOP BOTTLES

SAN RAFAEL   **SAN RAFAEL**   **FRESNO**   **MERCED**   **MADERA**
B B   **B B**
On Base   On Base

Borello Bros. also had a soda plant in Fresno, Merced and Madera. The plant manager was Andrew Borello. Partners were Allaria in Merced and Porter in Madera.

Buffalo Bottling Works box in Newall Snyder collection

BORELLO BROS.
SELTZER BOTTLES

John Louder collection

NOTICE OF MAUFACTURING, BOTTLING & SELLING

August 17, 1901

THE UNDERSIGNED BROS.
I, HAVE TO STATE THAT HE IS ENGAGED IN MANUFACTURING, BOTTLING AND SELLING

Soda Waters, Mineral Waters, Ginger Ale

and other beverages, in bottles, siphons, kegs and boxes, in the town of San Rafael, Marin County, State of California, with his principal place of business situated in the town of San Rafael, Marin County, State of California, and the following is a description of names and marks branded, stamped and blown upon the bottles, siphons, kegs, and boxes used by him in said business as afore said, to wit:

Soda bottles marked on side of bottle "M. Petersen, Marin Soda Works, San Rafael;" Marin Soda Works, San Rafael; "Soda Bottles also marked "M. B. W." on bottom of bottle; soda boxes marked "Marin Soda Works, M. Petersen, proprietor, corner D and First streets, San Rafael; copper kegs and tanks marked "M. Petersen."

                Borello Bros.

SAN ANSELMO BOTTLING COMPANY
SAN ANSELMO

- 1900 – 1910   William Petersen
- 1910 – 1912   F. M. Weinbeck
- 1912 – 1920   John B. Sutton & Francis Mulhern

A third San Rafael company, the San Anselmo Soda Works, was located in 1907 on D Street between Fourth and Fifth streets. Its proprietor was F. M. Weinbeck. In 1912, John B. Sutton and T. Francis Mulhern purchased the company and promoted their product with the motto "the best for the money."

Marin Journal
January 4, 1912

> T. F. Mulhern and J. B. Sutton have purchased the San Anselmo soda works, and the Peterson bottling works in San Rafael.

Marin County Tocsin
June 15, 1912

> Robt. Farrell of Bolinas has purchased the interest of J. B. Sutton in the San Anselmo Bottling Co. Mr. Farrell is well known in this city and many friends wish him success in his new venture. The firm now consists of Fred Weinbeck and Robt. Farrell.

SAN ANSELMO HUTCHINSON BOTTLES

FACE

**SAN ANSELMO
BOTTLING CO.
SAN ANSELMO, CAL.**

REVERSE

BLANK

FACE

**SAN ANSELMO
BOTTLING CO.
SAN RAFAEL, CAL.**

REVERSE

BLANK

SAN ANSELMO CROWN TOP BOTTLES

FACE

SAN ANSELMO
BOTTLING CO.
SAN RAFAEL, CAL.

REVERSE

BLANK

FACE

SAN ANSELMO
BOTTLING
CO.
SAN RAFAEL,
CAL.

REVERSE

BLANK

SAN ANSELMO SELTZER BOTTLES

FACE
(Acid Etched)

SAN ANSELMO
Bottling Co.
San Rafael, Cal.

John Louder Seltzer Bottle

FACE

SAN ANSELMO
BOTTLING CO.
SAN RAFAEL, CAL.

John Louder Seltzer Bottle

MEYER'S BEVERAGE COMPANY
"It's the water from the slopes of Mt. Tamalpais."

- Edmond Meyer

Marin Journal, October 6, 1921 - Agreement of Sale—Economic Machinery Co. to Meyer Soda Water Co. —one world labeler improved for $867.50, payable $86.75 per month.

Sausalito News, May 5, 1923 – Notice by Edmund Meyer that Meyer's Soda Water Co. is operating under the laws of California and all bottles are marked Bubbles, Meyer's Soda Water Co., Marin Bottling Works, E. Malz and Klammer & Malz and Martin Peterson. Bill of Sale Emil Malz to Edmund Meyer, soda water business consisting of machinery, bottles, etc.

His first drink in Marin was "Myers' Vitamin B Sparkling Water" announcing that it was "Recommended by Leading Physicians" and "Contains not less than 600 I. U. Vitamin B."

He established the Coca Cola distributorship in Marin County, and then purchasing the Coca Cola distributorship in Sebastopol he created an empire in the soda business. He moved the Sebastopol Coca Cola franchise to Santa Rosa in the late 1930's. The Marin plant was located on Second & Irwin Street.

FOR A REFINED TASTE
St. Francis Orange Dry
(Pure Orange Juice Carbonated)
An Appetizer and a Finishing Touch
for the Holiday Table
Bottled by the
ST. FRANCIS GINGER ALE COMPANY
Meyer's Soda Water Co.
San Rafael   *Distributors*   San Francisco

Sausalito News
December 7, 1928

MEYER'S PAPER LABEL SODAS

MARK HOPKINS
Lime Rickey
Meyer's Beverage Company
San Francisco - San Rafael

ST. FRANCIS
Pale Ginger Ale
Meyer's Specialty Beverage Co.
San Francisco, Calif.

MEYER'S BOTTLING CO.
Painted Labels

FACE
Meyer's
Trade mark

REVERSE

ARTIFICIALLY FLAVORED
AND COLORED
FRUIT ACID ADDED
BOTTLED BY
MEYER'S BOTTLING CO.
SAN RAFAEL, CALIF.

Bottle came in variety of flavors; Orange, Root Beer, Cherry, Cola, Grape, Lemon etc.

Meyer's bottle opener

MEYER'S BOTTLING CO.
Painted Labels

Meyer's used this particular painted label on numerous crown top bottles in the 1950's in various colors.

John Louder collection

MEYER'S BOTTLING CO.
Painted Labels

FACE
Meyer's
It's the Water
From the slopes of Mt. Tamalpais
QUALITY
BEVERAGES
ONE FULL QUART
MEYER BOTTLING CO.
SAN RAFAEL CALIFORNIA
CARBONATED WATER SUGAR ADDED OR ARTI-
FICIAL FLAVOR OTHER INGREDIENTS ON CROWN

FACE   REVERSE

REVERSE
IT'S THE WATER
FROM THE SLOPES OF
MT. TAMALPAIS
QUALITY BEVERAGES
IN STERILIZED
BOTTLES

MEYER'S EMBOSSED CROWN TOPS

FACE

Meyer's

REVERSE

Meyer's

BOTTOM

FACE

Meyer's
Special

REVERSE

Meyer's
Special

Bottles came in variety of flavors; Orange, Root Beer, Cherry, Cola, Grape, Lemon etc.

MEYER'S QUART BOTTLES

John Louder collection

MEYER'S SELTZER BOTTLES

FACE

MEYER'S
SODA WATER CO.
S AN RAFAEL
&
SAN FRANCISCO
NET CONTENTS 1 QUART

FACE

MEYER'S
SODA WATER CO.
MILL VALLEY
&
SAN FRANCISCO
CONTENTS 32 FL. OZ.

John Louder Seltzer Bottle

MEYER'S SELTZER BOTTLES

FACE

**MEYER'S SODA WATER CO.
SAN RAFAEL
CAL.**

FACE

**MEYERS COCA COLA
SAN RAFAEL & SANTA ROSA**

John Louder Seltzer Bottles

MEYER'S SELTZER BOTTLES

FACE

**MEYER BOTTLING CO.
(Emblem)
NET CONTENTS 1 QUART
SAN FRANCISCO – SAN RAFAEL**

FACE

**MEYER BOTTLING CO.
M. B. C.
SAN FRANCISCO – SAN RAFAEL**

John Louder collection

MEYER SELTZER BOTTLE

FACE

**TAMALPAIS
SPARKLING WATER
(Native scene)
MEYER BOTTLING CO.
SAN FRANCISCO – SAN RAFAEL
NET CONTENTS 1 QUART**

Tamalpais bottle Richard Siri collection

Tamalpais Mineral Water label Newall Snyder collection

MEYER'S COCA COLA
SAN RAFAEL, SEBASTOPOL & SANTA ROSA

WILL PAY 25c

FOR EACH EMPTY SIPHON BOTTLE RETURNED TO

Meyer Soda Water Co.

OR

Coca Cola Bottling Co.

Second and Irwin Streets — San Rafael

June 30, 1930
Sausalito News

WANTED!

Boys and Girls

To Pick Up Empty

SELTZER WATER BOTTLES

Return Them to Your Grocer or to

Meyer's Soda Water Co.
San Rafael, California

AND RECEIVE 25c FOR EACH BOTTLE

July 4, 1930
Sausalito News

> Agreement of Sale—Economic Machinery Co. to Meyer Soda Water Co.—one world labeler improved for $867.50, payable $86.75 per month.

Marin Journal
October 6, 1921

DOCUMENT FILED IN THE RECORDERS OFFICE

> Notnce — By Edmund Meyer, that Meyer's Soda Water Co. is operating under the laws of California and all bottles are marked Bubbles, Meyer's Soda Water Co., Marin Bottling Works, E. Malz and Klammer & Malz and Martin Peterson.
>
> Bill of Sale — Emil Malz to Edmund Meyer, soda water business consisting of machinery, bottles, etc.

Sausalito News
May 5, 1923

ADVERTISEMENTS

BOTTLES
OF EVERY DESCRIPTION FOR
Your Home Requirements

You are always welcome to visit our plant.

Meyer Soda Water Co.
SAN RAFAEL
Second and Irwin Sts.   Phone San Rafael 858

Sausalito News
November 23, 1928

Mill Valley Record
April 28, 1958

MEYER'S COCA COLA

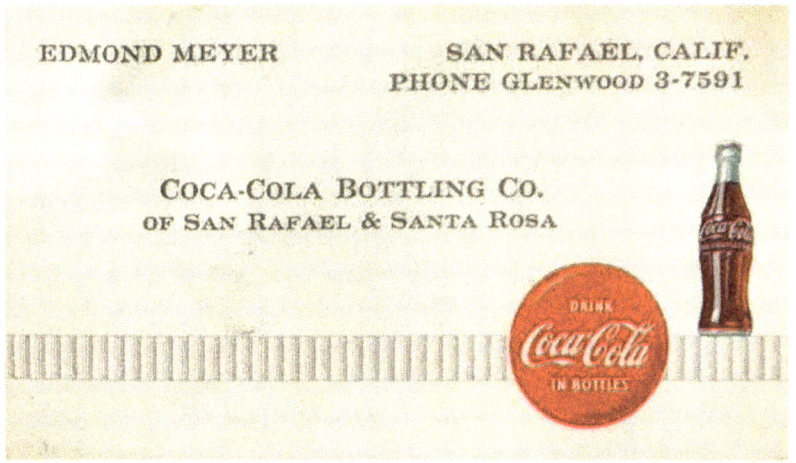

Business card Neal Austinson collection

Mill Valley Record
August 26, 1943

MASON & COMPANY
SAUSALITO

Sausalito Historical Society

View of Mason American Distillery plant of Sausalito that also manufactured soda water.

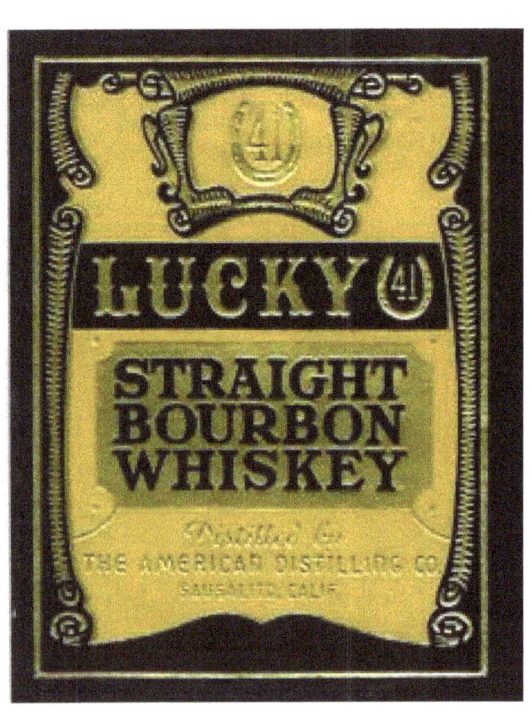

SHERIFF'S SALE

Order of Sale and Decree of Foreclosure and Sale.

THE SAUSALITO LAND AND FERRY COMPANY

Plaintiff,

vs.

THE MASON MALT WHISKEY DISTILLING COMPANY, SAMUEL L. LESYNSKY, JOHN MASON,

Defendants.

No. 1564

UNDER and by virtue of an Order of Sale and Decree of Foreclosure and Sale issued out of the Superior Court of the County of Marin, of the State of California, on the 21st day of April, A. D. 1896, in the above-entitled action, wherein the Sausalito Land and Ferry Company, the above-named plaintiff, obtained a Judgment and Decree of Foreclosure and Sale against the Mason Malt Whiskey Distilling Company, Samuel L. Lesynsky, John Mason, defendants, on the 18th day of December, A. D. 1895, which said judgment is recorded in Judgment Book 4 of said Court, at page 283, etc., I am commanded to sell all that certain lot, piece or parcels of land, situate, lying and being in the Township of Sausalito, County of Marin, State of California, and bounded and described as follows:

Beginning at a point on the southeasterly line of Alamo street, distant south 50 degrees 30 minutes west, one hundred and fifty (150) feet from its intersection with the southwesterly line of Bolinas street, as laid out and delineated upon a map entitled "official map of lands of Sausalito Land and Ferry Company," etc., filed in the office of the County Recorder of Marin County aforesaid, on the 26th day of April, 1869; running thence south 50 degrees 30 minutes west along said southeasterly line of Alamo street fifty (50) feet; thence at right angles south 39 degrees 30 minutes east and parallel with Bolinas street, two hundred and forty (240) feet, to the northwesterly line of Wateree street; thence at right angles north 50 degrees 30 minutes east along said northwesterly line of Wateree street one hundred (100) feet to a point on said line of said street distant one hundred (100) feet south 50 degrees 30 minutes west from its intersection with the said southwesterly line of Bolinas street; thence at right angles north 39 degrees 30 minutes west and parallel with Bolinas street one hundred and twenty (120) feet; thence at right angles south 50 degrees 30 minutes west and parallel with Wateree street fifty (50) feet; thence at right angles north 39 degrees 30 minutes west and parallel with Bolinas street one hundred and twenty (120) feet to said southeasterly line of Alamo street and the point of beginning. Being lots six (6), fifteen (15) and sixteen (16) in block seventy-two (72) as laid out and delineated upon the official map aforesaid.

Public Notice

is hereby given that on

Monday, the 18th Day of May A. D., 1896,

at one o'clock p. m. of that day, in front of the Court House door of the County of Marin, I will, in obedience to said Order of Sale and Decree of Foreclosure and Sale sell the above described property, or so much thereof as may be necessary to satisfy said judgment, with interest and costs, etc., to the highest bidder for gold coin of the United States.

San Rafael, April 22nd, 1896.

HENRY HARRISON,
Sheriff.

By JNO. HANNON,
Under Sheriff.

Marin County Tocsin
April 25, 1896

PIRATING RIVALS LABELS

FIRM IS CHARGED WITH PIRACY OF RIVALS' LABELS

Bottlers' Protective Association Sues to Prevent Use of Receptacles by Mason Soda Works.

Suit to prevent the use of bottles and cases bearing a manufacturer's label by firms other than those originally using them was filed yesterday by the Bottlers' Protective Association and Isaac H. Spiro against the Mason Soda Works.

Spiro appears in the case as a representative of the ninety or more firms that compose the association. These firms each have a distinct mark or sign, which appears on all bottles and cases containing their goods. It is claimed that the Mason Soda Works has made it a habit to collect these receptacles from customers after the contents have been used and has refilled them with goods of Mason manufacture.

A temporary restraining order and injunction is asked to protect the receptacles bearing the labels of manufacturers.

San Francisco Call
September 19, 1906

SMELL MINUS—MONEY PLUS

Aroma Surrounding Pine Station to be Bottled by Mason Distillery

All the commuters in Southern Marin will be pleased to know that the aroma that filled the air surrounding Pine station is going to be bottled and turned into money. The fact is that the Mason Distillery, the concern that has been making molasses into denatured alcohol at Pine, has awakened to the truth that it has been allowing a good source of revenue to escape in that odor that has been so apparent to all travelers and now it will be turned into a by-product called potash and used as a fertilizer. 'Tis true that every cloud has a silver lining, so we wish the distillery people a fortune from that smell.—Ex.

Marin County Tocsin
February 16, 1916

MANY MOTIONS AND CASES ARE GIVEN ATTENTION BY JUDGE ZOOK

Monday, Tuesday and Wednesday were taken up with the case of Geo. Lezinsky vs. John Mason. This is a case where Lezinsky claims a third interest in the distillery at Sausalito under an old agreement. Judge Seawell of Sonoma county presided and the case is to be continued and arguments heard March 7th at Santa Rosa.

On Thursday Judge Zook had returned and heard the case of R. H. Howe vs. M. A. Glamstead. This was put over until March 23.

Marin County Tocsin
February 19, 1916

THANKS BE!

Commuters, and all who travel to and from San Francisco over the Northwestern Pacific will be overjoyed to hear that John Mason of the Mason Distillery at Sausalito announced that after December 1st, the smell that all health authorities, civic bodies and private individuals have tried to abate for several years will be no more. Mr. Mason says that instead of dumping the refuse into the bay and thereby creating the offensive odor, it will be shipped eo eastern potash manufacturers and made into war materials.

Marin County Tocsin
October 28, 1916

More than fifty men were thrown out of employment one day last week when the Mason Distillery at Sausalito suddenly closed its doors as the result of criticism by commuters of the odor emanating from the refuse which flows from the plant into the bay. It is said that the owners of the plant are negotiating for the removal of the manufactury to Contra Costa county.

Marin Journal
November 1, 1917

Chance for a $500,000 Plant for Sausalito

Would Employ 500 Men—The Mason Distillery Has An Offer to Locate at Antioch

Through the opposition of certain well-meaning but ill-advised people, Marin county is about to lose the benefit of a large industry already established here, and also a much larger one for which plans had already been made and part of the work inaugurated.

The Mason Brothers of Sausalito, with the co-operation of Mr. John Buck and the Matson people, had succeeded in building up a large plant for the distilling of alcohol from molasses which heretofore had been allowed to run to waste in the Hawaiian Islands and which is brought to Sausalito in ships of the Matson line.

From a small concern of one still, under the direction of these men the plant has grown to be one of the largest in the world and the factory is now supplying all of the large powder making manufacturers with denatured alcohol, which is most essential to the manufacture of high grade explosives, and of course, very important in this war crisis.

The chemists in the employ of the company have discovered a process of further utilizing the waste product of molasses for the manufacture of potash, acetone, sulphate of amonia, acetic acid and several other chemicals. All of these chemicals have heretofore been largely manufactured in Germany and everyone well knows how essential they are to the manufacture of explosives, as well as useful in various other arts and crafts.

The company had plans drawn for the construction of buildings and supplying an equipment which would cost half a million dollars and which would employ at least 500 men, and when completed would absolutely take care of the waste product which at times runs into the bay. Material had been ordered and ground had been broken, old buildings removed and grading done.

A large number of workmen were employed when, without consulting with the managers of the Matson Company, and on complaint of three or four people living in the neighborhood, are attempting by every means in their power to secure the aid of the county authorities to cause the distillery to close down.

This has discouraged the promoters and they immediately ordered all new work destroyed and are making arrangements for erecting this new establishment in Antioch, Contra Costa county, where ninety acres of land are offered to them at a minimum price and quantities of fresh water will be available at only the cost of pumping it from the river to the plant.

It seems a pity that Marin county should lose this manufactory which would bring a number of highly trained men into the community and mean the disbursement of large sums of money monthly for wages, material and other necessities.

We think it behooves our Chamber of Commerce, both of Sausalito, Mill Valley and San Rafael, to make every effort to adjust this difficulty, especially as we understand the new enterprise will take care of the waste molasses which gives offense to the nostrils of some of the travelers on the train.—Independent.

Sausalito News
November 3, 1917

FEDERAL OFFICIALS VISIT MASON PLANT

Demonstrations of how alcohol could be illegally diverted from the Mason Distillery were made by federal prohibition agents in an official visit Thursday to the million dollar plant. The demonstrations were carried out as part of the hearing now under way to determine if the Mason By-Products company shall be granted a permit to continue operations. This was the first time since the beginning of the controversy that an actual visit to the plant has been made.

Clarence A. Linn, legal adviser to the prohibition department, who is prosecuting the case for the government, George Hurlburt, intelligence unit investigator, Theodore Roche, attorney for the defense, and Mason distillery officials made the trip across the bay.

Seeking to prove that there has been a shortage of alcohol in the huge plant and that the right to manufacture should be taken away, the government men traced over a series of pipe lines and valves running from the stills to the denaturing vats. The government is attempting to show that illegal withdrawals were made through an ingenious operation of these valves, according to Linn.

The case will probably be completed in a few days, when the defense will have finished its testimony.

Sausalito News
July 10, 1926

MASON PLANT CLEARED BY GOVERNMENT

Officers State Plant Has Been Operated 100 Per Cent According to Law

Mason's distillery, which has been under investigation by the prohibition forces for the last few months, has been proven clear of all charges against it, according to John Mason. It was alleged that alcohol manufactured in the plant was insufficiently denatured and that through secret pipes it was being removed. A complete examination of the plant and a hearing of the charges, however, proved to the government's satisfaction that the plant has been run 100 per cent in accordance with the law.

"It is unreasonable to suppose," Mr. Mason stated, "that we would even think of taking chances of having this $2,000,000 plant confiscated in order to sell our alcohol illegally. We are constantly rejecting offers of large sums for alcohol in which the denaturant has been left out, or to smuggle out alcohol. Moreover, we must be constantly on guard not to sell our alcohol to persons who will use it illegally.

"We are under a constant and great responsibility here, and we take every measure in our power to follow the letter of the law."

The charges made by the Board of Health against the Mason plant because of the odor, were dismissed after the board had investigated and found that everything was in a sanitary condition.

Alcohol, according to Mr. Mason, is one of the most necessary products to the industrial world. During the war, the government deprived its battleships of fuel oil, so that the distillery might continue to manufacture the alcohol which is all-important in ammunition making, one and one-half pounds of alcohol being required to make one pound of powder. As the distillery must be under an officer during war times, Mr. Mason received the rank of captain in the reserve army of the United States so that he might caary on the operations at the plant during the late war. This alone shows the government's confidence in his ability.

The Mason plant is now the largest alcohol plant in the world, the one at Baltimore having formerly ranked first. With the new buildings and equipment, however, the Sausalito distillery is now at the head.

Mr. Mason plans to beautify the front of the huge plant as soon as the highway is repaired and culverts put in so that the water may be drained under the highway, by planting a lawn next to the fence and forming a gravel stretch between this and the highway, where automobiles may park.

Sausalito News
October 16, 1930

Ernest Hopkins, in Examiner, Tells of Making of Dry Ice

One of the most interesting of the recent descriptive articles on Marin county, written by Ernest J. Hopkins in his "The Californian" column in the Examiner, was an account of the Industrial Solvents Corporation plant here in which he said:

SAUSALITO, Nov. 25.—Turning molasses into ice may strike you as a peculiar and unlikely proposition. But they do turn molasses into ice, and excellent ice it makes—ice that is so confoundedly cold that it burns your fingers like an electric shock when you pick it up. Ice that declines to melt as long as you keep air away from it, and melting, leaves no trace of moisture behind.

For that matter they could turn raisins into ice, or grapejuice, grainmash, or anything else that ferments. This modern discovery of "dry ice" bids fair to revolutionize the entire field of cold storage. To California, with its huge fresh-fruit output and its great distances from world-markets by land or sea, "dry ice" if you ask me is going to mean millions of dollars a year.

* * *

Every northbound motorist knows one place where, of late they are developing this by-product.

The place is, first and foremost, an alcohol factory.

Traditionally known as the Mason Distillery, it is properly entitled the Industrial Solvents Corporation; Walter E. Buck is president; John Buck vice president, and O. P. Heitmann is the manager. Fred King, as bonded warehouse agent, supervises the supply. This is the big and fragrant factory that stands beside the highway just at the northern end of Sausalito.

* * *

It happens that this Sausalito plant is the second largest alcohol manfactory in the United States—the largest one being in Pennsylvania.

A morbid interest attaches to the fact that, whereas this plant could turn out 5,000,000 gallons of 190-proof alcohol a year, it is permitted by the Federal government to turn out only 1,680,000 gallons for 1928; also that, whereas an investment of $600,000 is primarily designed for taking the poisons out of alcohol and making it pure, the product has to be artificially impurified again before it can be sold.

But then, as every one knows, alcohol is made nowadays only for such uses as quick-drying paint.

* * *

At the final stage in its career through the great vats, and endless coils of copper pipe and towering stills within this plant, the alcohol bubbles up into a series of glass enclosures like big percolator-tops. And here it is marvellously pure. The gauges showed they were making 192-proof stuff—200-proof being perfectly pure and of course theoretical.

To begin with, this alcohol was molasses. Next time you see an oil-tanker heading into this harbor on a return trip from Honolulu, Java or Cuba, if it is riding low it is carrying molasses. Tankers go out with fuel oil or gasoline—come back with molasses, which is the waste from sugar making and cheap. The molasses comes over to Sausalito in barges, and is simply pumped into the vats.

* * *

They put yeast into it, and the mischief begins. Here is where the "dry ice" comes in. When fermenting stuff bubbles, it gives off carbon dioxide. Heavier than air, that gas spills off down a pipeway is pumped into iron drums under 2,000 pounds of pressure. It becomes liquid inside these drums.

Now, if you turn on the strong little faucets and let that pressed gas suddenly expand, it solidifies at a temperature of 130 degrees below zero. That's mighty cold.

* * *

Funny, gritty stuff it is, rough and burning to the touch. Let it stand and it simply, in time disappears.

Load a refrigerator car with that "dry ice" instead of water ice, and—first—you have a temperature of 130 below instead of 30 above. Second, you have no deterioration from water dripping or rust. Third, the quantity to be carried is only one-fifth or one-fourth as heavy or as bulky. The problem is to keep it sufficiently airtight—but just now, for the ice cream trade, this "dry ice" is sent out in ordinary paper cartons, and lasts for days.

* * *

As to the fermented molasses, it goes on its way through various stills, getting purer as it goes. The last stills in the process are as high as a two-story house.

This plant has one storage tank that holds 1,000,000 gallons; smaller ones aggregating 200,000. The alcohol—properly impurified, depurified, anyhow spoiled—goes out in 50-gallon drums.

Once, when the Raisin Growers' Association had a tremendous carry-over, this plant bought the whole crop and made alcohol out of the raisins. The stacks solidified and had to be blasted loose with dynamite. In two weeks a tanker is going to bring in 14,000 tons of molasses.

And that's the way fifty or sixty Sausalito workers support their families. The "dry ice" output averages 40,000 to 50,000 pounds a month, now; as to alcohol, in September they had 10,000 50-gallon drums in the warehouse and now they're gone. Lots of quick-drying paint is used in this country.

Sausalito News
December 14, 1928

NOTICE OF SALE OF PARTNERSHIP PERSONAL PROPERTY

WHEREAS, on the 5th. day of December, 1928, there existed a partnership between De Witt C. Mason and John Mason, doing business under the firm name and style of Mason & Co.

AND WHEREAS, said John Mason, did on said 5th. day of December, 1928, by an instrument in writing elect to and did dissolve said partnership according to law and said partnership being dissolved and terminated ever since said 5th. day of December, 1928;

AND WHEREAS, ever since said dissolution and termination of said partnership on said 5th. day of December, 1928, all the personal property belonging to said partnership has been in the possession and under the control and is now in the possession and under the control of said De Witt C. Mason, as surviving partner of said Mason & Co.

AND WHEREAS, ever since said dissolution and termination of said partnership on said 5th. day of December, 1928, said De Witt C. Mason, has acted in liquidation of the affairs of said partnership;

AND WHEREAS, it is for the best interest of said partnership, and said De Witt C. Mason and said John Mason, in liquidation of the affairs of said partnership and the rights and interest of said De Witt C. Mason and said John Mason, to sell at Public Auction to the best and highest bidder all the personal property belonging to said partnership.

Notice is therefore given to all persons, that said De Witt C. Mason, as liquidating partner of said partnership, will on Tuesday, the 13th. day of August, 1929, beginning at the hour of one o'clock p. m. of said day, at and on the premises known and designated as 358 Litho Street, in the Town of Sausalito, County of Marin, State of California, sell at Public Auction to the highest and best bidder the following personal property of said partnership, to wit:

Notice is therefore given to all persons, that said De Witt C. Mason, as liquidating partner of said partnership, will on Tuesday, the 13th. day of August, 1929, beginning at the hour of one o'clock p. m. of said day, at and on the premises known and designated as 358 Litho Street, in the Town of Sausalito, County of Marin, State of California, sell at Public Auction to the highest and best bidder the following personal property of said partnership, to wit:

One Reo Truck, Engine Number 99059—One Dodge Truck, Engine Number 1073-043 — One old Reo Truck, Engine Number 23176—One old De Martini Truck, Engine Number 23176—One old Clydesdale Truck, Engine Number 7826On—One Water Filter—One Water Still—One Steam Boiler—One Bottle Washer, Complete—One Shield Carbinator—One Double Head Shield Automatic Bottling Machine—One Labeling Machine—about 20 soda tanks—about 1,000 cases of Soda Water Bottles—about 200 Syphon Bottles—Lot of Demijons—lot of Extracts—3 large crocks—Lot of Bottles and Flasks—Lot of used 5 gallon cans—about 100 Shipping Cases—One Office Safe and all other personal property of said partnership now in and upon said premises, at which may be upon said premises on the day of sale, whether mentioned herein or not, and also a lot of containers now in the possession of former customers of said Mason & Co.

All the personal property herein mentioned except said containers will be upon premises on said day of sale and any person may examine the same or may at any time after the date hereof during the day time and up to time of sale, examine all said property now upon said premises.

Terms of Sale: Cash, ten per cent of the selling price on the fall of the hammer, balance upon delivery of a bill of sale. Dated, August 2nd. 1929.

DE WITT C. MASON,

Liquidating partner of the dissolved partnership of De Witt C. Mason and John Mason, doing business under the firm name and style of Mason & Co.

Sausalito News
August 1, 1930

Clint Mason Still in Hot Water; Now Lawyer Louis Pistolesi Sues for Fees

Unkind fate still dogs the footsteps of DeWitt Clinton Mason, more generally known as Clint Mason, recent defendant in a prohibition case, and less recent defendant in a prosecution for scuttling a salmon boat. Louis C. Pistolesi, former counsel for Mason, has filed suit for the collection of attorney's fees, amounting in all to $3500.

Rumor states that Pistolesi, who has been Mason's attorney for some time, has also been unable, for some time, to collect certain fees due him for his valiant services in many cases in which Mason was co-defendant. Six days after the murder of Melvin Sturtevant, for which Vincent "Peg-Leg" Lucich was convicted, and in which Mason was allegedly involved, Pistolesi was discharged as Mason's counsel, and replaced by R. L. McWilliams.

Now this, since Pistolesi had defended Mason in several cases, and since he had yet to collect his fees, seemed frank ingratitude. Pistolesi had defended Mason when he was involved with Sturtevant in the case of the alleged stealing of 45 tierces of salmon from the Northern California Fisheries boat, "The Ideal". Mason and his son were found not guilty; but Sturtevant was found guilty. Pistolesi also represented Mason in the suit against him by John Mason, his brother, for an accounting following the dissolution of their partnership in the manufacture of industrial alcohol at the well-known local distillery.

This case was decided Wednesday by Judge Edward I. Butler, against Clint. Butler's opinion awarded $71,685.57 to John, which is an increase of over $24,000 over the amount awarded by Referee Paul Helmore.

And now Pistolesi has filed suit for collection of fees from Clint and Harvey Mason, his son, for $1000, and against Mason and the Progressive Mercantile Corporation for $2500.

Sausalito News
August 9, 1929

MASON & CO. HUTCHINSON SODA BOTTLE

FACE

MASON & CO.
SAUSALITO

REVERSE

BLANK

BOTTOM
Some are blank
others have

M
M M
M

Ebay listing

MASON CROWN TOP SODA BOTTLES

MASON
SODA WORKS
SAUSALITO

MASON
SAUSALITO
GINGERALE

Mason Sausalito Gingerale bottle Dan Brown collection.

MASON CROWN TOP SODA BOTTLES

FACE

MASON & CO.
SAUSALITO, CAL.

REVERSE

BLANK

FACE

DRINK
MASON'S
SODA
ITS GOOD

REVERSE

RETURN TO
MASON & CO.
SAUSALITO, CAL.

MASON SELTZER BOTTLES

FACE

**MASON
SODA WATER
SAUSALITO**

John Louder seltzer bottle

FACE

**SAUSALITO
SODA WORKS**

MASON SELTZER BOTTLES

FACE

**MASON
SODA WORKS
SAUSALITO**

John Louder seltzer bottle

SAN RAFAEL & SAN FRANCISCO

FORTUNA PURE MOUNTAIN WATER

At this time there is no information regarding Fortuna Water.
John Louder bottle

RECOMMENDED BOOKS

GRACE BROS. BREWERIES, HISTORY & MEMORABILIA
SANTA ROSA – LOS ANGELES – SACRAMENTO - FRESNO
John C. Burton
Aperitifs Publishing
Santa Rosa, Ca.
johncburton@msn.com

BOTTLE, TOKENS & HISTORY OF SONOMA COUNTY
John C. Burton
Aperitifs Publishing
Santa Rosa, Ca.
johncburton@msn.com

LAKE, NAPA, SONOMA, MENDOCINO, SOLANO,
MARIN, and HUMBOLDT MINERAL # HOT SPRINGS
John C. Burton & John Louder
Aperitifs Publishing
Santa Rosa, Ca.
johncburton@msn.com

CALIFORNIA'S BEST
OLD WEST ART AND ANTIQUES
Brad & Brian Witherell
Schiffer Publishing Company
4880 Lower Valley Road
Atglen, PA. 19310

HERE'S TO BEERS
Byron & Vicky Martin
8400 Darby Avenue
Northridge, CA 91324

CALIFORNIA HUTCHINSON TYPE SODA BOTTLES
Peck & Audie Markota
Self-Publish
Out of Print

UNITED STATES BEER CAN GUIDE
BEER CAN COLLECTORS OF AMERICA
747 Merus Ct.
Fenton Mo. 63026

UNITED STATES BEER CANS
With OPENING INSTRUCTIONS
Kevin C. Lilek
Lilek Publishing
2828 Breckenridge Circle
Aurora, IL. 60504

THE NEVADA BOTTLE BOOK
Fred Holabird & Jack Haddock
14040 Perlite Drive
Reno, Nevada 89511

BOTTLES & EXTRAS
Federation of Historical Bottle Collectors
101 Crawford Street Studio 1A
Huston, TX 77002
emeyer@fohbc.org

AMERICAN BREWERIES
Donald Bull, Manfred Friedrich, Robert Gottschalk
Bull Works Printing Co.
20 Fairway Drive
Stamford, Ct. 06903
203-969-1925

SPRINGS OF CALIFORNIA
By Gerald A. Waring
1915
Book located at Santa Rosa Library Annex

AUCTIONS

AMERICAN BOTTLE AUCTIONS
(Jeff Wichmann)
915 28th Street
Sacramento, CA 95816
916-443-3210
www.americanbottle.com

WITHERELL AUCTIONS
Brian Witherell
300 – 20TH Street
Sacramento Ca. 95811
916-446-6490

SHOW CHAIR PERSONS

NORTHWESTERN BOTTLE COLLECTORS ASSOCIATION
Santa Rosa, Ca.
Lou & Leisa Lambert
707-823-8845
nbca@comcast.net

GOLDEN GATE CHAPTER BOTTLE COLLECTORS
East Bay Area
Gary & Darla Antone
925- 373-6578
PACKRAT49ER@NETSCAPE.NET

CHICO BOTTLE SHOW CHAIRMAN
Randy Taylor
P. O. Box 1065
Chico, Ca 95927
530-518-7369
RTJARGUY@aol.com

RENO BOTTLE CLUB
Marty Hall & Helene Walker
775-335-9467
775-345-0171
rosemuley@att.net

49er CALIFORNIA CHAPTER
BEER CAN COLLECTORS OF AMERICA
Corry Weidman-Sibell
antque1plus@gmail.com

AMERICAN BREWERIANA ASSOCIATION
P. O. Box 269
Manitowish Waters, WI 54545
715-604-2774

JUST FOR OPENERS
John Standley
P. O. Box 51008
Durham. NC 27717

www.ingramcontent.com/pod-product-compliance
Lightning Source LLC
Chambersburg PA
CBHW061149070526
44584CB00034B/4467